T0240122

When 5G Meets Industry 4.0

Xiwen Wang • Longxiang Gao

When 5G Meets Industry 4.0

Xiwen Wang
Beijing Huaxia Institute of Industrial
Network Intelligent Technology Research
Beijing, China

Longxiang Gao
School of Information Technology
Deakin University
Burwood, VIC, Australia

ISBN 978-981-15-6731-5   ISBN 978-981-15-6732-2   (eBook)
https://doi.org/10.1007/978-981-15-6732-2

Image source: https://stock.adobe.com/de/images/abstract-city/15498053?prev_url=detail

This Springer imprint is published by the registered company Springer Nature Singapore Pte Ltd.
The registered company address is: 152 Beach Road, #21-01/04 Gateway East, Singapore 189721,
Singapore

Preface

With the development of global mobile communication technology, there is an intergenerational change cycle every decade. Since the 1980s, mobile communications have experienced changes from 1G to 4G at a rate of one generation per decade. Nowadays, the intergenerational upgrade in the past 10 years is coming again. 5G is opening a new era of large-bandwidth Internet and multi-connection, low-latency Internet of Everything.

1G was originated in 1984 and pioneered the era of mobile communications based on analog cellular technology. In 1990, 2G led the world into the digital communications era with two technical standards, namely GSM and CDMAOne. In 2000, the International Telecommunications Union (ITU) identified WCDMA, CDMA2000, TD-SCDMA, and WiMAX as the four wireless communication interface standards for 3G. In 2010, with the development of wireless communication standards, TDD-LTE and FDD-LTE, 4G technologies become matured and were commercialized on a large scale. With rapid development of the digital economy and mobile Internet, large-scale increase in smartphones and information consumption made mobile communication technology more prominent in people's daily life and social development. Previously, 4G did change our lives and provided people's basic requirements for video calling, short video sharing, car navigation, and video streaming. In the future, however, each family will have dozens of intelligent terminals on average and there will be millions of intelligent terminals connected to the network per square kilometer, and 4G will not be able to achieve that. In addition, the industry, transportation, medical, and other sectors are eager for low-delay connection, where 4G cannot provide due to its natural limitations.

5G technology has emerged as the times require, in order to cope with the explosive growth of mobile data traffic, massive device connections, and various emerging new services and scenarios with ultra-low latency in the future. 5G R&D has instantly become an important task for telecom operators and equipment manufacturers in major countries and organizations. From smart phones and smart watches to smart homes and smart grids, "smart" has become synonymous with innovation. In order to adapt to the rapid development of new technologies, the industrial field also refers to "smart"İ as describing intelligent factories and

v

intelligent manufacturing. As a result, the call for a new round of industrial revolution is getting higher and higher.

Germany's National Academy of Sciences and Engineering (Acatech) issued a report in April 2013 titled "Securing the Future of German Manufacturing Industry-Recommendations for Implementing the Strategic Initiative INDUSTRIE 4.0. (Final report of the Industrie 4.0 Working Group)." It hopes to use artificial intelligence to connect between information and the physical reality society and achieve the full integration of production processes and management processes. As a result, smart factories are realized, smart manufacturing is performed, and smart products are produced. Compared with the traditional manufacturing industry, the future smart manufacturing industry represented by smart factories is an ideal production system that can intelligently judge product attributes, production costs, production time, logistics management, safety, reliability, and sustainability to optimize product customization for each customer. 5G technology is the standard of the next generation mobile Internet. It not only enriches user experience, but also provides technical support for artificial intelligence–based applications, such as smart manufacturing, smart healthcare, smart government, smart cities, and driverless cars. As a result, 5G is regarded as the "infrastructure" of industrial Internet and artificial intelligence, therefore, both China and the USA are striving to become the leader of 5G to lead this new generation of international mobile communication international standards. As trade tensions between China and the USA continue to escalate, where products ranging from soybeans to mobile phones and automobiles may be affected, the key technology of 5G may actually be the cause of trade wars between the world's top two economies.

In short, 5G not only transforms society but also the international order. This book aims to describe 5G scenarios, changes, and values; explain the standards, technologies, and development directions behind 5G; and explore new models, new formats, and new trends of 5G-based artificial intelligence to readers.

Beijing, China   Xiwen Wang

Melbourne, Australia   Longxiang Gao
May 2020

Contents

1   The Development of 5G 1
   1.1   Generations of Mobile Communication System 2
   1.2   The Characteristics of 5G 5
   1.3   5G vs 4G 9
   1.4   What Changes 5G Can Bring? 12
   References 16

2   5G Technology System 17
   2.1   5G Standards 17
   2.2   Network Deployment of 5G 26
      2.2.1   Millimeter Wave 26
      2.2.2   Micro Cell Station 27
      2.2.3   Massive MIMO 30
      2.2.4   Device-to-Device 31
   2.3   Flexibility in 5G 33
      2.3.1   Edge Computing 34
      2.3.2   Network Slicing 38
   References 41

3   Development of Industry 4.0 43
   3.1   The Challenges of Traditional Industrialization 43
   3.2   Traditional Industrialization and AI 45
   3.3   Algorithm: The Core of the Industry 4.0 50
   3.4   Break the Impossible Trinity of Manufacturing Industry: The
      Vision of Industry 4.0 55
   3.5   ICT Technology: The Key to Industry 4.0 58
      3.5.1   What Is the CPS? 58
      3.5.2   Industry 4.0 Improves CPS from 3C to 6C 60
      3.5.3   CPS Realizes the Integration of Production Process
         and Information System 61
   3.6   Big Data: The Main Production Element of Industry 4.0 62
      3.6.1   An Era of Data Explosion 62

3.6.2   Increasing Data on Manufacturing 63
3.6.3   Big Data Provide Basis for Mass Customization 65
3.7   Intelligent Robots: The Main Force of Industrial 4.0 68
3.7.1   Why Do We Need a Large Number of Industrial
Robots for Intelligent Manufacturing? 68
3.7.2   Scenarios for Industrial Robots 71
3.7.3   The Latest Application Cases of Industrial Robots in
Intelligent Manufacturing 73
Reference 74

4 5G Communication Technology in Industry 4.0 75
4.1   eMBB to Achieve Industrial Internet 75
4.2   mMTC to Realize Digital Twin 84
4.3   uRLLC to Achieve Concurrent Manufacturing 86
4.4   5G LANs 90

5 5G in Real Industrial Scenarios 97
5.1   Perceivable Production Process 99
5.2   Early Warning Equipment Status 104
5.3   Improved Product Quality 110
5.4   Enhanced Man-Machine Cooperative 114

**6 Postscript: Industrial 5G: Open Intelligent Manufacturing
New Era** 119

Acronyms

3GPP	3rd Generation Partnership Project
AGV	Automated Guided Vehicles
AI	Artificial Intelligence
AMPS	Advanced Mobile Phone Service
APM	Asset Performance Management
AR	Augmented Reality
BBU	Building Baseband Unit
CA	Carrier Aggregation
CAD	Computer Aided Design
CAM	Computer Aided Manufacturing
CBRS	Citizens Broadband Radio Service
CM	Concurrent Manufacturing
CMM	Collaborative Manufacturing Model
CPS	Cyber Physical Systems
D2D	Device-to-Device
EDGE	Enhanced Data rates for GSM Evolution
eMBB	enhanced Mobile Broad Band
ERP	Enterprise Resource Planning
ETSI	European Telecommunications Standards Institute
FCC	Federal Communications Commission
FMS	Flexible Manufacturing System
GPRS	General Packet Radio Services
GSM	Global System for Mobile Communications
IMT	International Mobile Telecommunications
IoTs	Internet of Things
IT	Information Technology
ITU-R	International Telecommunication Union Radiocommunication Bureau
ITU	International Telecommunication Union
LTE	Long Term Evolution
MES	Manufacturing Execution System
MIIT	Ministry of Industry and Information Technology

MIMO	Multi-Input Multiple-Output
mMTC	massive Machine Type Communications
NFC	Near Field Communication
NFV	Network Function Virtualization
NIT	Network Information Technology
NR	New Radio
NSA	Non-Standalone
NSF	National Science Foundation
OEE	Overall Equipment Effectiveness
OT	Operational Technology
PLC	Programmable Logic Controller
PLM	Product Lifecycle Management
PPS	Production Planning System
QAM	Quadrature Amplitude Modulation
RAMI	Reference Architectural Model Industrie
RRU	Remote Radio Unit
RSPG	Radio Spectrum Policy Group
RTLS	Real-Time Locating System
SAS	Shared Access System
SDL	Supplemental Downlink
SDN	Software Defined Networks
TACS	Total Access Communications System
UAV	Unmanned Aerial Vehicle
UDN	Ultra-Dense Network
UMTS	Universal Mobile Telecommunications Service
uRLLC	ultra-Reliable Low-Latency Communication
V2I	Vehicle-to-Infrastructure
V2V	Vehicle-to-Vehicle
VR	Virtual Reality
WCDMA	Wideband Code Division Multiple Access
WRC	World Radiocommunication Conference

Chapter 1
The Development of 5G

This chapter mainly introduces the origin of 5G, what are its characteristics, and what are the performance improvements relative to 4G. 4G era is always "online," where train and hotel can view stock information, contact business, and enjoy videos or music, while 5G era is always "present," where it not only changes people's life but also the economy and society.

The essence of mobile communications is to achieve the transmission of information. Firstly, mobile communication transmits information by encoding the contents of voice, text, image, video, etc., and the core components such as optical module and baseband chips are utilized to convert them into radio wave signals. It is then transmitted by the core devices of base station, such as the RF antenna, and later accessed the core network via transmission media such as fiber-optic cables to achieve ultra-long-distance transmission. Finally, signals are decoded at the receiving end to restore the original information and present to the receiver.

In the process of communication, it is crucial to ensure the demand of high-speed rate, large bandwidth, and low delay. In addition, because massive data is converted into the optical wave transmission, it needs to make sure the transmission signals are not lost, do not interfere with each other, and finally are delivered to the receiver accurately. All links require a large amount of professional mobile communication technology and infrastructure construction investment. The development of mobile communication is to constantly make update iteration of software and hardware around these links.

Since the 1980s, each new generation of mobile communication technology is emerged every decade in the world, promoting the rapid innovation of ICT industry and the prosperity and development of economic society. 5G has both function and technology upgrade compared to the previous generation, namely, 4G mobile communications. It is the fifth generation (5G) upgrade of the global mobile communication industry. With a brand-new base station system and network architecture, 5G will not only provide a much faster speed, millisecond-level ultra-

© Springer Nature Singapore Pte Ltd. 2020
X. Wang, L. Gao, *When 5G Meets Industry 4.0*,
https://doi.org/10.1007/978-981-15-6732-2_1

low delay, and 100 billion-level network connection ability but also open a new era of extensive interconnection and deep human-computer interaction.

1.1   Generations of Mobile Communication System

The "G" of 5G refers to the first letter of the word "generation." In other words, 5G is the fifth generation of mobile communication system. Each generation will be more advanced than the previous generation, and its functional performance is further improved, as shown in Fig. 1.1.

As early as 1820, Hans Christian Oersted, the Danish physicist, found that when an electric current passed through a wire, the magnetic needle placed near it would deflect [1]. Many years later, in 1887, Heinrich Rudolf Hertz, a young German physicist, revealed the great truth of the existence of electromagnetic waves through experiments, opening up an infinite and broad prospect for human beings to use radio waves [2]. This is where communication technology traces its origins – "Hz" (hertz) – and it has also become a frequency unit in the International System of Units.

The application of electromagnetic waves in the communication field has its uniqueness and inevitability. Firstly, electromagnetic wave is a kind of energy, which has the possibility of generating and absorbing. It has extremely high matching with the transmission and reception of information. Secondly, it is generally believed that the speed of light is the fastest speed in the universe, and the propagation of electromagnetic waves in a vacuum speed is the speed of light, which allows

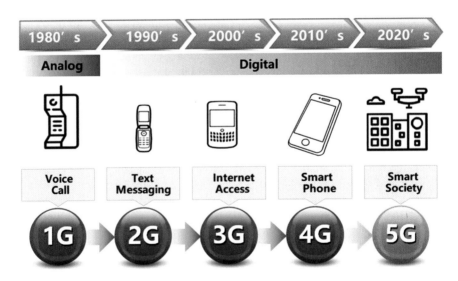

Fig. 1.1 Mobile communication appears a new generation of technology every 10 years

electromagnetic waves to meet the speed requirements of information transmission to the greatest extent. On the basis of matching theory and reality, the first generation of mobile communication system was born in Chicago, USA, in 1986, which is the 1G network.

The reason why each generation of mobile communication technology can achieve faster speeds, lower delays, and more stable transmission is through the evolution of technology and the adjustment of the architecture by improving the bandwidth of available frequency bands and the transmission efficiency of existing frequency bands (Table 1.1).

- 1G (voice call): The 1G mobile network was put into use in the early 1980s. It has voice communication and limited data transmission capabilities (early capacity is about 2.4 Kbps). 1G network uses analog signals, such as AMPS (Advanced Mobile Phone Service) and TACS (Total Access Communications System) standards, to "pass" cellular users among distributed base station (hosted on the base station tower).
- 2G (message passing): In the 1990s, 2G mobile network spawned the first batch of digitally encrypted telecommunications, which improved voice quality, data security, and data capacity and provided limited data capabilities through the use of GSM (Global System for Mobile Communications) standard circuit switching. In the late 1990s, 2.5G and 2.75G technologies used the GPRS (General Packet Radio Services) and EDGE (Enhanced Data rates for GSM Evolution) standards to increase the data transmission rate (up to 200 Kbps). Later 2G iterations introduced data transmission through packet switching, which provides a stepping-stone for 3G technology.
- 3G (multimedia, text, Internet): In the late 1990s and early twenty-first century, 3G networks were introduced with a faster data transmission speed by fully transitioning to data packet switching. Some of these voice circuit switching were already 2G standards, which made data streaming possible. The first commercial 3G service was introduced in 2003, including mobile Internet access, fixed wireless access, and video calling. 3G network now uses standards such as UMTS (Universal Mobile Telecommunications Service) and WCDMA (Wideband Code Division Multiple Access), which increase the data speed to 1 Gbps in the stationary f state and 350 Kbps or higher in the mobile state.
- 4G (real-time data: car navigation, video sharing): 4G network services were launched in 2008, making full use of all IP networking and relying entirely on packet switching. Its data transmission speed is ten times compared with 3G network. The 4G network's large bandwidth advantage and extremely fast network speed have improved the quality of video data. The popularity of LTE networks sets communication standards for mobile devices and data transmission. LTE is constantly evolving, and the 12th version is currently being released. The speed of "LTE-A" can reach to 300 Mbps.

From analog communication to digital communication and from text transmission, image transmission to video transmission, mobile communication technology has greatly changed people's lifestyle. The first four generations of mobile commu-

Table 1.1 Communication generation comparison before 5G

	First generation mobile communication (1G)	Second generation mobile communication (2G)	Third generation mobile communications (3G)	Fourth generation mobile communication (4G)
Start time	1980s	1990s	Early twenty-first century	2010s
Commercial time (wordwide)	In 1978, Bell Labs developed the Advanced Mobile Phone Service (AMPS) for the first time	In 1989, Europe entered commercial applications with GSM (Global System for Mobile Communications) as a standard	In October 2001, Japan's NTT DoCoMo operator first launched WCDMA service in the world	2010 World Mobile Communications Conference focuses on Long-Term Evolution (LTE)
Commercial time (China)	In 1987, China began to deploy 1G networks	In 1993, Jiaxing GSM network officially became China's first digital mobile communication network	In 2009, 3G licenses were officially issued to the three major operators, and China entered the 3G era	In December 2013, the Ministry of Industry and Information Technology in China officially issued 4G licenses for the three major operators, and China entered the 4G era
Representative companies	Motorola	Nokia	Apple and Samsung	Apple, Samsung, and Huawei
Main feature	Analog signal transmission, voice call	Digital signal transmission, voice call, SMS, simple low-speed data service	Can transmit sound and data information at the same time, provide high-quality multimedia services	Fast transfer of data, audio, video, and images
Disadvantage	Low voice quality, unstable signal, and poor anti-interference	Limited data transmission capacity and weak communication encryption	Limited user capacity, low transmission rate, and inconsistent transmission standards	Too many frequency bands are used worldwide and do not support IoT transmission

nication network technology are only focused on mobile communication, while 5G also includes many application scenarios such as the industrial Internet and artificial intelligence.

5G (Smart Internet of Everything), faced with such a complex and changing application environment, is not just a simple upgrade of mobile communication technology, but an innovative change to the overall base station construction and network architecture. Unlike the past 2G to 4G era, which focused on mobility and transmission rates, 5G must consider not only enhancing broadband but also the large-scale connectivity and ultra-low latency required by the Internet of Everything, as well as diversified future needs, key technologies, evolution paths, and many other dimensions.

Since 2009, China's Huawei company has begun the prospective layout early research on 5G-related technologies. After 10 years' research and development, 3GPP has completed the full version of 5G standards and completed the submission of the IMT (International Mobile Telecommunications)-2020 standard (Table 1.2).

In 2017, through the joint efforts of the 5G industry, 5G-related standards, key technologies, the 5G industrial environment have made breakthrough progress. From 2018 to 2019, 5G has entered the field test and pre-commercial stage, where large-scale field tests were conducted extensively, and standards and technologies were further improved. It is expected that 5G will start commercialization on a large scale between 2020 and 2021. The main goal of 5G network construction at this stage is to off-load the pressure of 4G networks and further increase the wireless network bandwidth. After 2022, 5G commercialization will continue to deepen and expand, the communications industry and vertical industries will merge across borders, and more new formats, models, and scenarios will emerge. In particular, 5G, as a key enabling technology and infrastructure in the future digital economy era, will strongly support the intelligent transformation of vertical industries, such as smart manufacturing, smart agriculture, smart healthcare, smart cities, smart environmental protection, and intelligent robot.

1.2 The Characteristics of 5G

5G, as the next hot spot in the mobile communication industry, plays a significant role in both the communications industry and the entire social economy.

The International Telecommunication Union Radiocommunication Bureau (ITU-R) defines three typical application scenarios of 5G as enhanced mobile broadband (eMBB), ultra-reliable low-latency communication (uRLLC), and massive machine-type communications (mMTC) (Fig. 1.2). Among them, eMBB is mainly for high-bandwidth demand services such as virtual reality (VR), augmented reality (AR), and online 4K video; mMTC is mainly for services with high connection density requirements such as smart cities and intelligent transportation; uRLLC is mainly for delay-sensitive services such as Internet of Vehicles, unmanned driving, and UAV (unmanned aerial vehicle).

Table 1.2 Important milestones of 5G development

Time	Milestone
2009	Huawei start its earlier studies conducted 5G related technologies
February 13	EU announces funding of 50 million euros to accelerate research and development of 5G mobile technology
May 13	South Korea's Samsung announces that it has successfully developed a 5G core chip
May 14	Japanese telecommunications operator, NTT DoCoMo, announces that it cooperates with manufacturers such as Ericsson, Nokia, and Samsung to develop 5G technology
October 15	ITU (International Telecommunication Union) officially names 5G technology as IMT-2020 and expects to complete standard development in 2020
	China, Europe, Japan, and South Korea's 5G Promotion Organization signs agreement, laying the foundation for a globally unified 5G standard
April 16	Huawei takes the leadership in completing the first phase of China's IMT-2020 (5G) promotion group (air interface key technology verification test)
July 16	Nokia collaborates with Canadian operator, Bell Canada, to complete Canada's first 5G network technology test
October 16	Qualcomm released the world's first 5G modem, Snapdragon X50 modem
November 16	The 3GPP (the 3rd Generation Partnership Project) has determined that the PolarCode scheme promoted by Huawei has become the standard coding scheme in the 5G control channel, the eMBB scenario. The LDPC scheme promoted by Qualcomm has become the uplink and downlink coding scheme of the data channel
April 17	KT and Verizon open the world's first 5G holographic video call
February 18	Vodafone and Huawei announce that the two companies have jointly completed the world's first 5G call test in Spain using the non-independent 3GPP 5G new wireless standard and Sub-6 GHz (low-frequency band)
Jun 18	The 5G NR standard SA (Standalone) solution was officially completed and released, which marked the official launch of the first truly complete international 5G standard
2019 ~	5G enters the field test and pre-commercial stage

Specifically, the concepts and definitions of the three typical application scenarios are as follows:

- eMBB, literally translated as "enhanced mobile broadband" which simply means "large bandwidth." It is a human-centered application scenario, which focuses on ultra-high data transmission rates and mobility assurance under wide coverage. In the next few years, user data traffic will continue to show explosive growth (average annual growth rate is as high as 47%), and the business type is also dominated by video (accounting for 78%). With the support of 5G, users can easily enjoy both 4K/8K video and VR/AR video online. The user experience rate can be increased to 1 Gbps, while 4G is up to 10 Mbps, and the peak rate can even reach 10 Gbps.
- uRLLC, literally translated as "ultra-reliable low-latency communication" which simply means "ultra-low latency." In future application scenarios, the connection

Enhanced Mobile Broadband

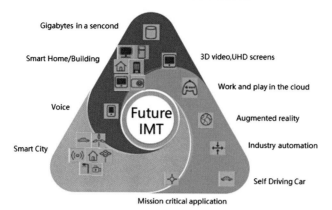

Fig. 1.2 ITU defines three major application scenarios for 5G in 2015

delay must reach the level of 1ms, and it must support high-reliability (99.999%) connections at high speeds (e.g., 500 km/h). This scenario is more oriented from industrial applications such as Internet of Vehicles, industrial control, and telemedicine.

- mMTC, literally translated as "massive machine-type communications" which simply means "large-scale connection." 5G's strong connectivity can quickly promote the deep integration of vertical industries (smart manufacturing, smart agriculture, smart cities, smart homes, environmental protection, environmental protection, etc.). In the era of the Internet of Everything, people's lifestyles will also undergo dramatical changes. In this scenario, the data transmission rate is low and insensitive to delay, and the connection will cover all aspects of the economy and society.

Therefore, the industry generally believes that with the advent of 5G, higher speeds, larger bandwidths, and lower delays become possible. Also some services that could not be achieved by the past mobile communication technology become possible, thereby realizing the large-scale innovation of business application. It is expected to further exploit the consumption potential and expand the total consumption. In addition, it also has a significant driving effect on equipment manufacturing and information service links. Looking to the future, 5G will become the key infrastructure for comprehensively building an intelligent transformation of the economy and society. From online to offline, from the consumer Internet to the industrial Internet, and from the emerging artificial intelligence industry to the intelligent upgrade of traditional industries, it will promote the development of the digital economy and the construction of a smart society to a new level.

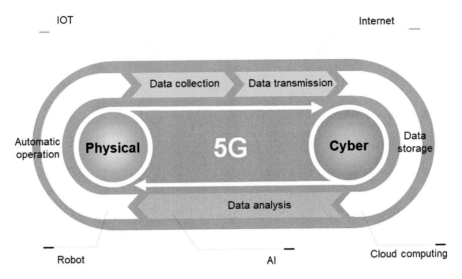

Fig. 1.3 Deep integration of IoT with big data and artificial intelligence

In fact, as early as 10 years ago, the Internet of Things had attracted much attention from all walks of life. Many people think that the Internet of Things is a way to completely change every aspect of one's life. The deep integration of the Internet of Things with other new technologies, such as big data and artificial intelligence, will form solutions for many platforms. Artificial intelligence will provide algorithms that analyze big data collected by IoT devices, identify various modes, and make intelligent predictions and decisions (Fig. 1.3). With the increase of the number of IoT devices and the amount of data generated, the large-scale connectivity of 5G enhanced networks is particularly important. 5G technology will achieve a wider network coverage, a more stable Internet connection, and a faster data transmission speed (from 1 Gbps to 10 Gbps in 4G). It will also allow more mobile devices to access the network at the same time, thereby realizing the real interconnection of everything.

5G has not only created tremendous opportunities for all walks of life but also laid the foundation for large-scale subversion. According to Ericsson's forecast [3], major 5G networks are expected to be deployed by 2020, and it is estimated that 4.1 billion IoT devices will use 5G communications worldwide by 2024.

As countries around the world vigorously promote Industry 4.0 and smart manufacturing, 5G will be more widely and deeply integrated into the industrial field. More 5G LAN wireless connections will appear in the factory workshop, which will promote continuous optimization of the network infrastructure, effectively improving the level of networked collaborative manufacturing and the quality of factory workshop. 5G technology can not only help the manufacturing industry to become more flexible and efficient in production operation but also improve safety and reduce maintenance costs and enable "smart factories." For example, manufacturing industry can improve self-optimization processes using remote control, monitoring, and dynamic configuration of industrial robots by through 5G mobile networks.

Medical systems need faster and more efficient networks to keep up with the large amounts of data it processes, from detailed patient information to clinical research and to high-resolution MRI and CT images. By introducing 5G technology into the medical industry, it will effectively meet requirements such as low latency, high-definition image quality, and high reliability and stability in the telemedicine process, promote the rapid popularization of telemedicine applications, and achieve the remote diagnosis, treatment, and consultation for patients (especially in remote areas). For example, 5G can enable monitoring devices (such as wearable technology) to have a longer battery life while sending patient health data to doctors in real time. On another side remote robotic surgery can also take advantage of 5G's low-latency and high-throughput communications to transmit high-definition image stream, so as to promote the precise realization of remote surgery.

5G will also subvert media and entertainment on multiple levels, including mobile media, mobile advertising, home broadband, and television. Due to the low latency of 5G, streaming video is unlikely to stop or freeze. On 5G networks, movie downloads will be reduced from an average of 7 min to just 6 s [4]. When browsing social media, playing games, streaming music, or even downloading movies and programs, 5G will save people an average of 23 hrs loading time per month. According to a study conducted by Ovum, an authoritative and neutral consulting company in the world's telecommunications industry, the global media industry will receive a staggering $76.5 billion in cumulative revenue through new services and applications enabled by 5G technology over the next decade.

Of course, 5G can achieve more than these scenarios.

1.3 5G vs 4G

In a sense, the International Telecommunication Union Radiocommunication Bureau (ITU-R) defines the "eMBB" scenario, one of the three typical application scenarios of 5G, that is actually the upgraded version of 4G mobile communication technology. It is just that the eMBB has once again improved the traffic and bandwidth processing capabilities of 5G mobile communication technology, supporting a higher and faster content transmission processing. The main application scenarios include video services such as 4K and 8K, voice, image, text processing, mobile app data, and other consumer-oriented services.

The charm of 5G is that while it uses eMBB technology as an upgraded version of the original 4G communication, it also introduces two other major scenarios, that is, mMTC for large-connected IoT applications and uRLLC for low-latency ultra-high-speed networks. Therefore, the key difference between 5G and previous

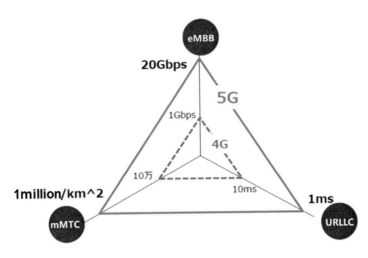

Fig. 1.4 Comparison of 5G and 4G

generations of mobile communication technologies is that the previous generations of networks are committed to improving the user experience for consumers (and to a certain extent for enterprise users), while 5G is targeted at industrial applications, which will be unprecedented to support a wider range of scenarios.

Therefore, in general, the key to distinguish 5G from 4G is not only its fast speed and wide bandwidth but also its more efficient and stable ubiquitous connection of the Internet of Things, as well as ultra-high-speed low-latency network transmission. Specific to the technical level, these differences can be compared from the three major application scenarios of 5G technology: eMBB, mMTC, and uRLLC (Fig. 1.4).

For mMTC, this technology is more inclined to the application of IoT scenarios. Although 4G communication technology supports the use of IoTs, it is unsuitable for large-scale applications, as 4G technology is more targeted at consumer applications such as mobile calls, video, and text transmission. 4G communication technology has defects in the IoT scene, but mMTC technology is specifically aimed at solving this defect of 4G technology. The requirements of IoT transmission communication technology are more important in the breadth and width of devices that can be connected. Although it may not require as much data transmission rate and bandwidth as consumers, it needs to maintain a large amount of data and support high concurrency, multi-channel data transmission and processing capabilities. This part of the business is another different area in communication capability compared from eMBB scenario. At present, smart cities and smart environmental protection are the most typical representatives of mMTC application scenarios.

The technology of uRRLC is mainly designed for areas that required a higher data transmission rate and sensitivity, such as autopilot and telemedicine surgery. Autopilot involves high-speed autonomous vehicles, and it must be able to quickly reflect the brakes and identify and avoid accidents. On another side, if the control sensitivity of surgical robot is not enough in telemedicine surgery, it may easily cause the patient's wound unable to heal, thereby failing to achieve high-precision minimally invasive surgery. In these scenarios, the required data transmission and processing capabilities are different from those in other fields, so the birth of communication technology uRRLC came into being.

From the perspective of communication, each of the three scenarios has its own strengths. As they are all communication technologies and the ultimate goal is signal processing and transmission, there are only technologies more suitable for a certain scenario among them, rather than saying that a certain technology must be used to solve a certain field. So for the industries that these technologies can really empower, it is more about which technology the players in this industry are better at using and which technology they tend to choose.

Of course, from the perspective of mobile communication technology development, the maturity of 5G technology and the industrial chain require a long-term process. It is expected that 4G will coexist and cooperate effectively with 5G for a long time. In the future, 5G will be combined with new capabilities and networks such as artificial intelligence, cloud computing, and the IoTs to achieve cross-border integration with vertical industries. It will also create new formats in the fields of power, logistics, banking, automotive, media, medicine, smart cities, etc. and open up a huge value growth space for the industry.

So how does 4G transition to 5G?

It is expected that from 2019 to 2020, in order to control costs and make a smooth transition to 5G, telecommunications operators first provide 5G ultra-high-speed services based on new frequency bands to large cities with high communication needs. 5G will initially be deployed in urban dense areas, with the goal of increasing user network speeds-enhancing mobile broadband scenarios. Currently, carrier aggregation (increasing bandwidth capacity) and network optimization technologies are used to increase network speed. The base station using the new radio (NR) will coexist with the 4G LTE base station and will operate in a non-standalone (NSA) mode. At this stage, it is mainly based on LTE and LTE-Advanced networking. According to the research progress of 3GPP [5], the uplink carrier aggregation (CA) and 256-QAM (quadrature amplitude modulation) are highly efficient. The 4G frequency band is adopted to provide services for smart phones. Non-independent networking will use existing 4G infrastructure to deploy 5G small base stations in areas with high service density. Although, non-independent networking can bring 5G to the market faster, it may be more suitable for local hot spot deployments rather than large-scale national deployments. In addition, the interoperability between non-independent networking and existing LTE networks is also very complicated. At present, different network strategies meet the needs of different operators, while operators also choose different network deployment paths at different points for their 5G deployment.

After 2020, with the continuous construction and use of 5G core networks, NR base stations with independent network (standalone) will begin operation, officially providing 5G services with ultra-high speed, large-scale connections, high relia-bility, and low latency. Independent networking will form a new network, including new base stations, backhaul links, and core networks. The advantage of independent networking is that it can form large economies of scale under the premise of providing high performance and avoid problems such as complex interoperability that may occur in the process of integration with LTE network. However, in the early stage of commercialization, the cost of independent networking is relatively high (Fig. 1.5).

1.4   What Changes 5G Can Bring?

As the fifth generation of mobile communication technology, 5G is fundamentally different from the previous four generations of mobile communication technolo-gies. The first generation is analog technology, the second generation realizes digital voice communication, the third generation is characterized by multimedia communication, and the fourth generation has greatly increased communication rate and made the entry of wireless broadband era. The previous four generations are just using single mobile communication technology, while 5G is the sum of the previous four generations of technology and adds high-frequency commu-nication technology, which makes 5G have higher communication peaks, lower latency, greater transmission capacity, and lower power consumption. Therefore, the commercialization of 5G will promote the upgrading of the value chain of the communications industry, as well promote the prosperity and development of the economy and society (Fig. 1.6).

The arrival of 5G era, as a new basic network facility, serves not only people but also things and society. 5G connectivity, from financial services to healthcare and retail, will drive the intelligent interconnection of everything. Driven by its three major characteristics, low latency, high bandwidth, and high-speed rate, it will boost the development of the Internet of Everything and bring unlimited business opportunities to the artificial intelligence industry. At the same time, whether it is home life, agriculture, forestry, aquaculture and construction industry, medical treatment, education, or disaster rescue and relief, 5G will give people huge changes from four aspects: ease, presence, sensitivity, and intelligence (Fig. 1.7).

Driven by the high speed of 5G, ultra-clear video develops rapidly, which is also one of the hot spots that major telecom operators are currently competing for. The application of ultra-clear video brings huge revenue to telecommunications operators. In terms of infotainment, 5G will promote the development of video and gaming applications in the direction of ultra-high-definition, 3D, and immersive experience and become an indispensable network support for new applications such as 8K ultra-high-definition video. In learning, people are able to deploy VR and AR technology into the virtual classroom, by wearing devices immersive involved in their favorite courses, and panoramic communicate with teachers and students

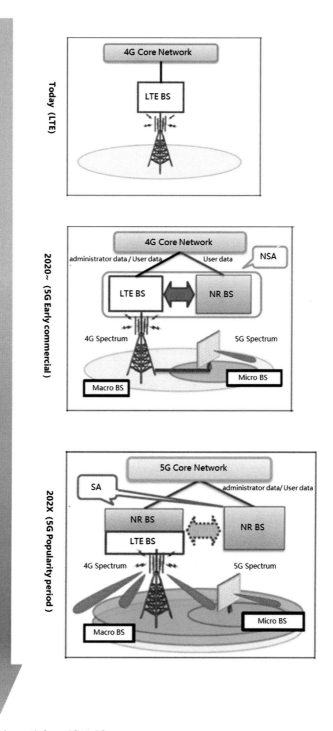

Fig. 1.5 Transition path from 4G to 5G

	1G (1980~)	2G (1993~)	3G (2001~)	3G IMT 3.5G (2006~)	3.9G (2010~)	4G IMT-Advanced (2014~)	5G (2020~)
Communication method	Different standards (Analog)	PDC(Japan) GSM(EUR) cdmaOne(NA)	W-CDMA CDMA2000	HSPA EV-DO	LTE	IMT-Advanced	5G
Rate	——	Several Kbps	384Kbps	14Mbps	100Mbps	1Gbps	20Gbps
Main services	📞	✉	🖥 (♪) 🎮		▶ ▶		🦾 🚗
Main object	people						Thing

Fig. 1.6   5G is essentially different from the previous four generations of communication technology

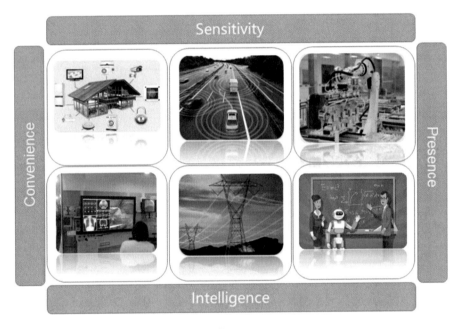

Fig. 1.7   5G gives people huge changes from four aspects

in the classroom. It is foreseeable that the arrival of 5G era will drive AR and VR technologies into mainstream technologies. In the past, the lack support of high-speed network technology, AR/VR technology has serious delays, which brings uncomfortable experiences such as dizziness. Once 5G solves the problems of its speed and delay, AR/VR will break through the limitation of bottleneck, make a qualitative leap, and have broad market prospects.

With the arrival of 5G, there are no restrictions on delays and speeds, and it has solved the particularly core mobile communication technology problems for the unmanned driving of connected cars. It is believed that in the near future, the popularity of driverless cars will be further promoted, and the vehicular network industry will develop in a blast.

The Internet of Vehicles is a subversive change in 5G and IoTs technology in the transportation industry. It provides people with integrated services by integrating related information such as people, cars, roads, and the surrounding environment. Relying on the advantages of low latency, high reliability, high speed, and security of 5G, it effectively improves the ability of timely and accurate collection, processing, dissemination, utilization, and security of Internet of Vehicles information. The interoperability and efficient coordination of information with road will help eliminate the security risks of the Internet of Vehicles and promote the rapid development of the Internet of Vehicles industry. For example, V2V (vehicle-to-vehicle) communication must take place in real time, because milliseconds can lead to call fails and fatal collisions. Achieving this high-speed interconnection requires vehicles to transfer large amounts of data between each other without any lag. 5G can achieve this goal through their reliability and low latency. In addition, 5G plays a key role in V2I (vehicle-to-infrastructure) communications. V2I communication connects vehicles with infrastructure such as traffic lights, bus stops, and even the highway itself. This can improve traffic flow, reduce external risk factors, increase vehicle response speed, and improve public transport efficiency.

As 5G technology continues to advance and is embedded into a large number of terminals, machines, and processes, wireless communications will have a transformative impact on various industries and regions and will lead a new era of innovation and economic development. In the future, the deep integration of 5G with cloud computing, big data, AI, VR/AR, and other technologies will connect people and everything and become a key infrastructure for digital transformation in all walks of life. On the one hand, 5G will provide users with more immersive business experiences such as ultra-HD video, next-generation social networks, and immersive games and promote the upgrade of human interaction methods. On the other hand, 5G will support massive machine communications, and the typical application scenarios represented by smart cities and smart homes will be deeply integrated with mobile communications. It is expected that hundreds of billions of devices will access 5G networks. More importantly, 5G will also explode vertical industry applications such as Internet of Vehicles, mobile medical care, industrial Internet, etc., with its excellent performance of ultra-high reliability and ultra-low latency (Fig. 1.8).

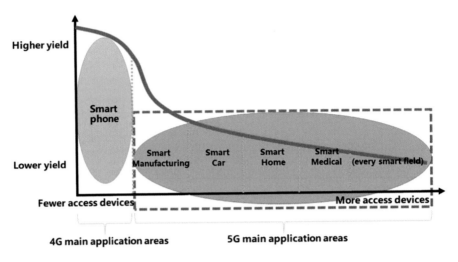

Fig. 1.8 Application directions of 5G

References

1. B. Weiss, "Selected scientific works of hans christian Ørsted. translated and edited by Karen Jelved, Andrew D. Jackson and Ole Knudsen. with an introduction by Andrew Wilson. Princeton: Princeton University Press 1998. 647 Seiten," *Berichte zur Wissenschaftsgeschichte*, vol. 23, no. 1, pp. 58–59, 2000.
2. A. N. D. Baird, R.I. Hughes, "Heinrich hertz: Classical physicist, modern philosopher," *Boston Studies in the Philosophy and History of Science*, vol. 198, 1998.
3. Ericsson, "5G estimated to reach 1.5 billion subscriptions in 2024 – ericsson mobility report." https://www.ericsson.com/en/press-releases/2018/11/5g-estimated-to-reach-1.5-billion-subscriptions-in-2024--ericsson-mobility-reportm. Accessed: 2020-03-31.
4. A. Moscaritolo, "5G will save you almost 24 hours of download time per month." https://au.pcmag.com/why-axis/59138/5g-will-save-you-almost-24-hours-of-download-time-per-month. Accessed: 2020-03-31.
5. 3GPP, "3GPP ran: Rel-12 and beyond." https://www.3gpp.org/news-events/1579-ran_rel12_and_beyond. Accessed: 2020-03-31.

Chapter 2
5G Technology System

This chapter mainly introduces the 5G standard and its technology system. With the development of the mobile Internet, people have higher and higher requirements for the communication experience in various application scenarios. They are hoping to obtain a good business experience in ultra-dense scenarios such as stadiums and concerts. The requirement of this seamless access at any time has transformed the focus of the network from limited coverage to limited capacity. For new services, the demand of mobile video services is becoming the mainstream. Different applications have different requirements for mobility, frequency, delay, and traffic. With the rapid development of 4K video and virtual reality services, the demand for bandwidth is also increasing. For example, the bandwidth demand of 1080p video service is only 4 Mbps, while the bandwidth of 4K video service has reached 20 Mbps, and the bandwidth demand of VR service will reach more than 100 Mbps in the future. Therefore, in the early stage of 5G deployment, especially in the aspect of enhanced mobile broadband, the performance improvement of core network and wireless access network is very important.

2.1 5G Standards

In a series of frequency allocation and identification, international organizations, regional telecommunication organizations, and national regulatory agencies need to coordinate the global uniform spectrum standards, which is one of the prerequisites for the successful deployment of 5G network. Only through coordinated allocation can radio interference at the border be minimized, international roaming is facilitated, and equipment costs are reduced.

Therefore, the overall coordination of the global unified spectrum is also the main goal of the ITU-R in the World Radiocommunication Conference (WRC) process.

© Springer Nature Singapore Pte Ltd. 2020
X. Wang, L. Gao, *When 5G Meets Industry 4.0*,
https://doi.org/10.1007/978-981-15-6732-2_2

The high-capacity deployment capabilities of 5G require more spectrum bandwidth, which increases the demand for spectrum. At present, the industry is working together to develop a path for the development of 5G spectrum. For example, the ITU is coordinating the international coordination of 5G mobile system to develop additional spectrum. The standardization department of ITU plays a key role in formulating the technical and architectural standards of 5G system.

Historically, the WRC carries out a major mobile communication spectrum allocation approximately every 8 years. In 1992, WRC-92 divided the 3G core frequency band, which became the basis for 3G development. In 2000, WRC-2000 divided 2.6 GHz frequency band, which is an important frequency band of 4G. In 2007, WRC-07 divided the 3.5 GHz frequency band and digital dividend frequency bands, which are the current hot spots for global 4G development. In 2015, WRC-15 allocated 470–694, 1427–1518, 3300–3400, 3600–3700, and 4800–4990 MHz to IMT for use in some regions or countries, which are important mid-band resources for 5G development. The 2015 Radiocommunication Assembly (RA-15) approved "IMT-2020" as the official 5G name. At this point, IMT-2020 forms a new IMT series with existing IMT-2000 (3G) and IMT-A (4G). This indicates that the frequency bands currently marked for IMT systems in the ITU Radio Regulations can be considered as the low- and medium-frequency bands for 5G systems (Fig. 2.1).

At the same time, in order to actively respond the rapid growth of data traffic in future mobile communications, WRC-15 conference identified WRC-19 agenda item 1.13. In accordance with Resolution 238 (WRC-15), it reviewed and determined the frequency band for the future development of IMT, including the possibility of making additional divisions for mobile services as a primary service.

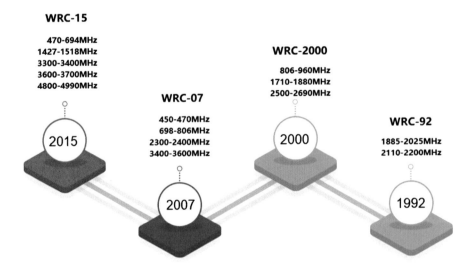

Fig. 2.1 WRC and IMT spectrum identifier

It also requested ITU-R to carry out research, including spectrum requirement research on the ground portion of IMT in the frequency range of 24.25–86 GHz, the eight mobile service bands (24.25–27.5, 37–40.5, 42.5–43.5, 45.5–47, 47.2–50.2, 50.4–52.6, 66–76, and 81–86 GHz), and three frequency bands that have not yet been allocated by the mobile service (31.8–33.4, 40.5–42.5, and 47–47.2 GHz).

The research content of this item includes three aspects: the prediction of spectrum demand, the research of candidate frequency band, and the analysis of interference coexistence between systems.

1. The prediction of spectrum demand is mainly to analyze the necessity of new spectrum. Specifically, the research of spectrum demand is based on historical data, integrating various influencing factors of future development, combining with the prediction trend of mobile communication data growth, considering the carrying capacity of specific technology system, analyzing the future frequency demand, and giving the total amount of spectrum needed in different stages as the basis of new spectrum.
2. The research of candidate frequency band is based on the research conclusion of spectrum requirements, and the appropriate target frequency band is selected and proposed. It is necessary to fully consider the comprehensive factors such as service allocation, mobile communication system requirements, equipment and device manufacturing capabilities, etc. and initially select a suitable target frequency band. Countries and standardization organizations propose preliminary candidate frequency bands based on the current situation of frequency used in their countries and regions.
3. The inter-system coexistence study mainly evaluates the availability of the selected target frequency band. According to the business division, system planning, and application status of the proposed candidate frequency band, and based on the existing business or system's technical characteristics, deployment scenarios, and other factors, research between the mobile communication system and other existing or planned systems (millimeter-wave frequency band mainly focuses on space business) can be conducted.

In the CPM19-1 meeting (the first WRC-19 preparatory group meeting after WRC15), it was determined that the ITU-R study group responsible for this item is the 5G millimeter baud working group (TG5/1), which was responsible for compatibility coexistence analysis and formed a CPM report to give recommendations for global 5G frequency planning. At the same time, it was further determined that ITU-R WP5D completed IMT spectrum demand prediction, ITU-R SG3 was responsible for the propagation model required by the coexistence research institute, and other ITU-R groups including SG4, SG5, SG6, and SG7 were responsible for providing TG51 with parameters and protection criteria of the original services on the relevant frequency band.

The main goal of the WRC-19 agenda item 1.13 is to seek global or regional coordinated millimeter-wave bands for 5G, which is an important basis for conducting 5G millimeter-wave research worldwide. Therefore, the research direction of this issue has an important impact on global 5G frequency planning. Most countries or regions will carry out planning based on the progress and results of the item. In a sense, if a country or region is to lead the development of global 5G spectrum, it is necessary to rely on item 1.13 to globalize national or regional perspectives through item research. For WRC-19, this process is currently at the stage of reaching consensus on IMT allocation and identification of large contiguous blocks of the world's coordinated radio spectrum that can obtain large bandwidths above 24 GHz, and these blocks have large bandwidths available. WRC-19's decision on this subject will be based on ITU's study of the broad sharing and compatibility of mobile and existing services in these and adjacent bands.

In addition to the International Telecommunication Union (ITU), the 3rd Generation Partnership Project (3GPP) is a recognized global mobile communications standardization organization, where it is also accelerating research on 5G NR frequency bands.

At the 71st RAN Plenary Session of 3GPP in March 2016, the research topic "Study on New Radio Access Technology" was adopted to study 5G-oriented NR access technologies. At present, according to the 3GPP 5G roadmap, the development of 5G NR standards based on deployment requirements is divided into two phases: The first phase of the standard was completed in June 2018 (Rel. 15) to meet the needs of early 5G network deployment before 2020. The phase 2 needs to consider compatibility with phase 1. It is planned to be completed at the end of 2019 (Rel.16) and submitted to ITU-R IMT-2020 as a formal 5G version (Fig. 2.2).

During the research phase of 5G NR, 3GPP carried out research on channel models above 6 GHz (3GPP TR 38.900), as well as studied and determined the requirements and scenarios of NR (3GPP TR 38.913). Based on this, it started the evaluation of NR technology solutions and proposed a series of NR access technology solutions to support Rel 15 standard formulation. The 75th Plenary Session of the 3GPP RAN held in March 2017 passed the conclusion of 5G NR access

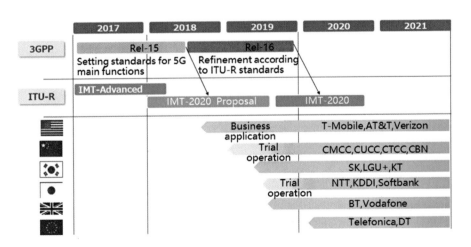

Fig. 2.2 Evolution of 5G standards

technology research project and officially launched the Rel 15 standard formulation work for 5G NR access technology. In the proposal for project establishment, the NR frequency band to be defined (including the new NR frequency band range and LTE re-cultivation frequency band) and the dual connection of NR and LTE or the frequency band combination of CA are listed and updated continuously according to the demand.

Telecommunications regulators in some European and American countries are considering using 700 MHz, 3.4, and 24 GHz bands for the initial deployment of 5G to meet its coverage and capacity requirements. Next, the licensing and use mode of 5G spectrum above 24 GHz will be considered. GSMA expects that the 3.3–3.8 GHz spectrum will be the basis for many of the original 5G services, particularly the provision of enhanced mobile broadband. This is because 3.4–3.6 GHz covers almost the whole world and it can well promote the economies of scale required for low-cost equipment. In addition, 3GPP is currently developing standard projects including OFDM-based scalable waveforms, a new flexible framework that supports lower latency and forward compatibility, and new advanced antenna technologies that utilize high-frequency bands.

At present, 5G standards are gradually taking shape. Global telecommunications operators have begun to increase and intensify network deployment. Base station construction and application trials are in full swing, and a new era of smart things is about to begin.

5G mobile technology continues to spread in various industries and processes with limited wireless popularization, and it will have a profound and lasting impact in a wide range of industries and regions.

Firstly, 5G technology will enable mobile technology to expand beyond consumer and enterprise services to industry applications, allowing people to interact with the world in an unprecedented way. As mentioned earlier, the technical specifications and functions of 5G will be completely different from previous generations of network technologies. A wide range of terminals will use multiple radio types to complete a range of diverse tasks.

Secondly, 5G standards will not only use licensed and unlicensed spectrum but also use shared spectrum to operate on both private and public networks. This high level of flexibility indicates that 5G will be able to handle an unprecedented number of industry use cases. For the mobile ecosystem to successfully penetrate these industries, the key is to gain a deep understanding of the different industries and use cases it addresses. The terminal life cycle of many industries will be as long as 10 years or more. Other industries may require dedicated networks or network requirements for specific spectrum segments.

5G will not only serve the mobile communication itself but also penetrate into various fields of the future society. The integration and innovation with the traditional manufacturing and service industries will promote the "smart+" new business form; change people's production, work, and lifestyle; and bring infinite vitality to the development of the global economy and society. Compared with previous generations of mobile communication systems, 5G needs to meet more diverse scenarios and extreme performance requirements. Therefore, spectrum resources are the key basic resources for the development and deployment of 5G network. To deploy a feasible 5G network, the selection and availability of spectrum is the most important factor, because spectrum determines the speed, quantity, and delay of data transmission in the future.

While using existing spectrum is efficient and flexible, developing new spectrum for 5G network is critical for future development. The "5G Vision" issued by the ITU-R defines that 5G systems will meet enhanced mobile broadband, massive inter-machine communication, ultra-high reliability and ultra-low time, and the three major application scenarios of extended communication.

In terms of system performance, 5G systems will have a peak rate of 10–20 Gbit/s and a user experience rate of 100 Mbit/s–1 Gbit/s. Compared with 4G systems, it will increase spectrum efficiency by three to five times, energy efficiency by a hundred times, and many key capability indicators such as mobility support at 500 km/h, air interface delay in 1 ms, connection density in 1 million/km^2, and traffic density in 10 Mbit/s/m^2.

Based on the above requirements of key performance indicator, in order to meet the application requirements of 5G systems in different scenarios, the candidate frequency bands of 5G systems need to be oriented to the full-frequency band layout, low-frequency and high-frequency bands, and planned as a whole to meet network requirements for capacity, coverage, and performance. Spectrum resources are a key factor driving 5G standards and industrial processes. In the process of finding new spectrum resources, the mobile communication industry will inevitably be subject to huge resistance from other industries. How to balance the development of mobile communications with satellite, national defense, scientific research, broadcasting, and other services is essential to provide resource support for 5G's future development.

Spectrum availability has the greatest impact on 5G development and plays a key role in 5G operations, development, and promotion. Countries are currently adopting two different approaches to deploy new spectrum with hundreds of MHz bandwidth for 5G to significantly improve its performance. One method focuses on the part of the spectrum below 6 GHz (low-frequency to middle-frequency spectrum, also known as "sub-6"), mainly in the 3–4 GHz frequency band, and the representative country is China. The second method focuses on the part of the spectrum ("high spectrum" or "mmWave") in the 24–300 GHz, and the representative country is the United States.

The low- and medium-frequency spectrum below 6 GHz can take into account the coverage and capacity of the 5G system. It is designed for the three major application scenarios of eMBB, mMTC, and uRLLC to build a 5G basic mobile communication network. The high-frequency spectrum above 6 GHz is mainly used to achieve the capacity enhancement of 5G network and realize the hot spot and fast experience for the scene of eMBB.

5G use cases may be satisfied by various spectrum frequencies. For example, low-frequency and short-range applications (suitable for dense urban areas) may be suitable for mmWave frequencies (above 24 GHz). Long-range and low-bandwidth applications (more suitable for rural areas) may be suitable for frequencies below

1 GHz. Although a lower frequency has better propagation characteristics for better coverage, a higher frequency can support higher bandwidths, which is due to the greater spectrum availability of the mmWave band.

There are three advantages of mmWave. Firstly, short wavelengths and narrow beams provide better resolution and security for data transmission and can transfer large amounts of data faster with minimal delay. Secondly, mmWave has more bandwidth available, which can increase data transmission speed and avoid congestion in low spectrum bands. Although the mmWave ecosystem requires a lot of infrastructure to build, it can get somewhere to transmit data at speeds up to 20 times the current 4G network speed. Finally, mmWave components are smaller than those used for lower-frequency bands, allowing for more compact deployment on wireless devices.

However, the challenge with mmWave technology is that the transmission distance is limited and it is easily blocked by obstacles such as walls, leaves, and the human body itself. This creates high infrastructure costs, as mmWave networks require dense base stations throughout the geographic area to ensure an uninterrupted connectivity.

Sub-6 can provide a wide area network coverage and has a lower risk of interruption compared to mmWave. This is because it has a longer wavelength and greater ability to penetrate obstacles. Compared to mmWave, it requires less capital expenditure and fewer base stations. Coupled with the ability to take advantage of existing 4G infrastructure, allowing the sub-6 spectrum to build a 5G ecosystem faster, it may become the dominant global frequency band that drives infrastructure and equipment deployment.

Spectrum, as a basic strategic resource for wireless communications, is critical to the development of the 5G industry. In order to guide the development of 5G industry and seize market opportunities, starting in 2016, major global countries or regions, including the United States, the European Union, South Korea, and Japan, have formulated 5G spectrum policies.

The United States achieves 5G high- and low-frequency spectrum layout

The US Federal Communications Commission (FCC) has opened spectrum resources for 5G technology in the high-, medium-, and low-frequency bands and summarized three main points.

1. Plan to enrich high-frequency resources. On July 14, 2016, the United States unanimously approved the rules and regulations for the use of spectrum above 24 GHz for wireless broadband services and planned a total of 10.85 GHz high-frequency spectrum for 5G wireless technology, including 28 GHz (27.5–28.35 GHz), 37 GHz (37–38.6 GHz), and 39 GHz (38.6–40 GHz) with 3.85 GHz licensed spectrum and 64–71 GHz with 7 GHz unlicensed spectrum. On November 16, 2017, the FCC released a new spectrum plan and approved the use of a total of 1700 MHz spectrum resources in the 24.25–24.45 GHz, 24.75–25.25 GHz, and 47.2–48.2 GHz frequency bands for the development of 5G services. So far, the FCC of the United States has planned a total of 12.55 GHz millimeter-wave spectrum resources.

2. Attach importance to the sharing of middle-frequency bands. In April 2015, FCC of the United States provided 150 MHz of spectrum for the Citizens Broadband Radio Service (CBRS) in the 3.5 GHz band (3550–3700 MHz), established a three-layer spectrum Shared Access System (SAS) regulatory model, and allowed for test. On the basis of protecting existing services, SAS has exerted market mechanisms and introduced public wireless broadband services. AT&T has officially proposed to the FCC special temporary authority to test 5G equipment in the 3.5 GHz band.

3. Release low-frequency resources. At the WRC-15 conference, the United States identified the 470–698 MHz digital dividend frequency band in phase 2 as IMT system by adding footnotes. In April 2017, the auction of 600 MHz frequency band was completed, and T-Mobile became the biggest winner and planned to use it for 5G deployment.

EU releases 5G spectrum strategy and strives to seize 5G deployment opportunities

On November 10, 2016, the European Commission's Radio Spectrum Policy Group (RSPG) released the European 5G spectrum strategy, which clearly stated that the 3400–3800 MHz band will be the main frequency band for 5G deployment in Europe before 2020 and 700–1000 MHz will be used for 5G wide coverage. In terms of millimeter-wave bands, it is clear that the 26 GHz (24.25–27.5 GHz) band will be used as the initial deployment for 5G high-frequency bands in Europe. Some of them will be used to meet the 5G market demand by 2020. In addition, the European Union will continue to study the 32 GHz (31.8–33.4 GHz), 40 GHz (40.5–43.5 GHz), and other high-frequency bands.

Japan releases radio policy report specifying 5G spectrum range

On July 15, 2016, the Ministry of Internal Affairs and Communications (MIC) of Japan released a radio policy report for 2020, identifying 5G candidate frequency bands as low frequencies include 3600–3800 and 4400–4900 MHz, and high frequencies include 27.5–29.5 GHz Bands and other WRC-19 research frequency bands. For 5G commercial use in 2020, Japan focuses on the 3600–3800 and 4400–4900 MHz frequency bands and the 27.5–29.5 GHz frequency bands.

South Korea changes C-band plan and clarifies 5G spectrum

On November 7, 2016, the Ministry of Science and ICT and Future Planning (MSIP) of South Korea announced that the 3.5 GHz (3400–3700 MHz) spectrum originally planned for 4G was converted to 5G. The recovered 3.5 GHz spectrum in 2017 will be used as 5G spectrum and re-licensed. During the 2018 PyeongChang Olympic Games in South Korea, three operators deployed 5G test networks in the 26.5–29.5 GHz band to demonstrate 5G services.

German 5G spectrum plans to cover four bands of high, middle, and low frequencies

Germany announced its national 5G strategy on July 13, 2017, by releasing more 5G spectrum plans, specifically involving four frequency bands. The 2 GHz band, that is, 1920–1980 and 2110–2170 MHz, are mainly used for 3G services in Germany. The current license will expire in 2025. After the expiry date, Germany plans to continue to use them as the 5G working frequency band. The 3.4–3.8 GHz frequency band is used for mobile communication. Germany has completed the auction in June 2015 for the 700 MHz frequency band, and the next step will continue to put 738–753 MHz into 5G as SDL (Supplemental Downlink). For the 26–28 GHz bands, unlike the European Union, Germany has decided to use the 28 GHz band as the 5G band, specifically 27.8285–28.4445 and 28.9485–29.4525 GHz. At the same time, Germany has not completely excluded the 26 GHz band and continued to use it as a research band.

The United Kingdom releases 5G spectrum plan

Ofcom's 5G spectrum planning report released in February 2017 [1] indicates that its 5G spectrum will be consistent with the European Union's RSPG, selecting 3.4–3.8 GHz, 24.25–27.5 GHz, and 700 MHz as the high-, middle-, and low-frequency spectra, respectively. At present, the United Kingdom has completed the clean-up work in 3.4–3.6 GHz frequency band and carried out the clean-up work in the 700 MHz frequency band.

On the whole, the global cognition of 5G's spectrum architecture is basically the same as that of coordinating spectrum resources in the high-, medium-, and low-frequency bands. In the future, 5G networks will be coordinated networking with high and low spectrum. The mid-band mainly refers to the C-band (3400–3800 MHz) which will be the core frequency band for global 5G deployment and the main coverage and capacity layer of the 5G network. The high-frequency bands 24.25–27.5, 28, and 40 GHz are 5G network ultra-large capacity layer, which are used to meet the needs of high-capacity and high-speed services. Below 1 GHz, such as 700 and 600 MHz, is the coverage layer of 5G networks, which mainly meets the needs of the wide area and deep indoor coverage.

China specifies the middle-frequency resources for 5G deployment

In order to adapt and promote the application and development of 5G systems in China, it released a frequency use plan for 5G systems in 3000–5000 MHz band at the end of 2017, which defined 3300–3400 MHz (indoor use in principle), 3400–3600 and 4800–5000 MHz as the working frequency band of 5G system and defined the middle-frequency resources for 5G deployment.

In terms of high-frequency bands, Chinese authorities also rely on the WRC-19 agenda item 1.13 research group, IMT2020 (5) promotion group, and other platforms to carry out related work. Relying on the WRC-19 agenda item 1.13 platform, the frequency authority leads the organization of relevant units to conduct 24.75–27.5 compatibility analysis of 5G systems and other services in the 24.75–27.5 and 37–42.5 GHz frequency bands. In June 2017, the Ministry of Industry and Information Technology (MIIT), China, conducted a public opinion collection on 24.75–27.5, 37–42.5 GHz, or other millimeter-wave bands for 5G systems. At the Asia-Pacific Telecommunity conference (APG19) held in July 2017, China stated its point of view to prioritize the study of candidate frequency bands in the 24.75–27.5 and 37–42.5 GHz bands. On July 3, 2017, MIIT newly added 4.8–5, 24.75–27.5, and 37–42.5 GHz frequency bands for China's 5G technology R&D tests.

2.2   Network Deployment of 5G

2.2.1   Millimeter Wave

Wireless communication is the use of electromagnetic waves, where their functional characteristics are determined by their frequency. Electromagnetic waves of different frequencies have different attributes and thus have different uses.

Radio waves are a type of electromagnetic waves, and their frequency resources are limited. In order to avoid interference and conflicts, it is usually necessary to divide different frequency bands and assign them to different objects and uses. The salient feature of electromagnetic waves is the higher the frequency, the shorter the wavelength, and the closer the straight line propagation (the worse the diffraction ability). The higher the frequency, the greater the attenuation in the propagation medium.

All along, mobile communication is mainly carried out by using middle frequency to ultra-high frequency. With the development of 1G, 2G, 3G, and 4G, the radio frequency used is getting higher and higher (Table 2.1). If mobile communication uses a high-frequency band, the transmission distance will be greatly shortened, and the coverage capability will be significantly weakened accordingly.

Millimeter waves operate at high frequencies between 30 and 300 GHz. There are many reasons why millimeter waves are so attractive (Fig. 2.3). First, millimeter waves with shorter wavelengths produce narrower beams, which provides better resolution and security for data transmission. It also can transmit a large amount of data, with a short delay. Secondly, if more millimeter-wave bandwidth is available, it not only improves the data transmission speed but also avoids the congestion in the low-frequency band (before studying the application of millimeter-wave frequencies in 5G, this band was mainly used in radar and satellite services). The 5G millimeter-wave ecosystem requires a large-scale infrastructure, and it can achieve data transmission speeds 20 times higher than 4G LTE networks. Finally, millimeter-wave components are smaller than components in the lower-frequency bands, so they can be deployed more compactly on wireless devices.

Although millimeter wave has many benefits, it faces various challenges. Although the characteristics of short wavelengths and narrow beams can improve

Table 2.1 Radio frequency of mobile communication is getting higher and higher

	First-generation mobile communication (1G)	Second-generation mobile communication (2G)	Third-generation mobile communication (3G)	Fourth-generation mobile communication (4G)
Start time	1980s	1990s	Early twenty-first century	2010s
Frequency band	300–3400 Hz	900–1800 MHz	1880–2145 MHz	1880–2665 MHz

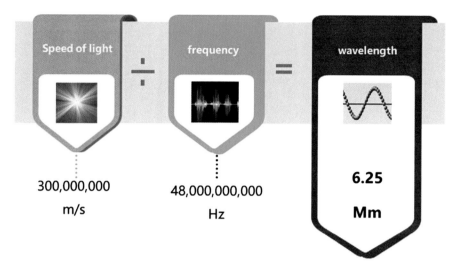

Fig. 2.3 Millimeter wave

resolution and transmission security, it limits the propagation distance. Because the millimeter-wave network needs to be spread throughout the entire area covered by the base station and maintain an uninterrupted connection, this will incur high infrastructure costs. Millimeter waves are easily blocked by obstacles such as walls, leaves, and the human body, which further exacerbates this challenge.

Mitigating physical challenges requires more new technologies and designs, such as massive MIMO (Multi-Input Multiple-Output). Massive MIMO is an antenna array that greatly expands the number of device connections and data throughput, enables the base station to accommodate more users' signals, and significantly increases the capacity of the network.

Obviously, the higher the frequency, the richer the frequency resources that can be used. The richer the frequency resources, the higher the transmission rate that can be achieved. As 4G data transmission capability cannot keep up with current requirements, 5G upgrade will solve the problems of speed and attenuation by deploying a network using millimeter wave. Of course, the number of 5G base stations required to cover the same area will also far exceed the requirement of 4G.

2.2.2   Micro Cell Station

5G is targeted at the three application scenarios of eMBB, mMTC, and uRLLC, which need to provide different network performance. On the wireless side, there are a large number of new technologies to support different application scenarios, but

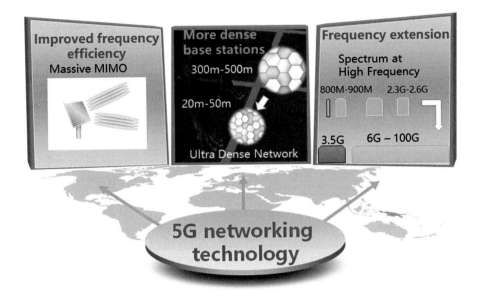

Fig. 2.4 Key technologies for new-generation mobile communication performance improvement

the transmission network side has limited hardware technology upgrades. Therefore it is necessary to reform the network architecture (Fig. 2.4).

Ultra-dense network (UDN) will be the main technical to meet 5G and future mobile data traffic requirements. UDN can achieve a higher-frequency reuse efficiency through more "dense" deployment of wireless network infrastructure, thereby achieving a hundred-fold increase in system capacity at local hot spots. With the increase of site density, users may have the same frequency interference from multiple dense neighboring cells, the switching will be too frequent when moving, and the user experience will drop dramatically. The Pre5G UDN solution can turn interference from multiple base stations into useful signals, and the service set is continuously updated as the cell moves. It is always keeping the user in the center of the cell, realizing cell virtualization, and achieving a consistent user experience. Interference management and suppression, cell virtualization technology, and small cell dynamic adjustment are important research directions for ultra-dense networking in the Pre5G UDN phase.

In 2020 and beyond, the popularity of ultra-HD, 3D, and immersive video will significantly increase data rates. A large amount of personal and office data are stored on cloud, and massive real-time data interaction needs to be comparable to the transmission rate of optical fiber. The communication experience requirements in complex and diverse scenarios are getting higher. In order to meet the needs of users to obtain a consistent business experience in the ultra-dense scenes of large gatherings, outdoor gatherings, and concerts, 5G wireless networks need to support 1000 times the capacity gain and 100 billion of high-capacity hot spots for this

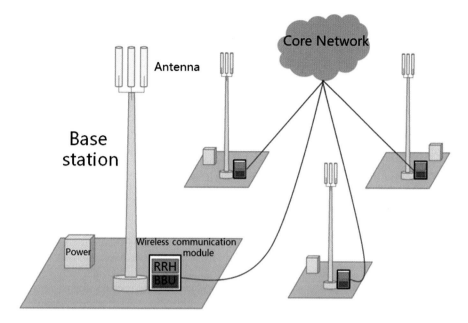

Fig. 2.5 Schematic diagram of the base station

scenario in the future. UDN can increase the system frequency reuse efficiency and network capacity by increasing the deployment density of base stations. It will become a key solution for hot spot high-capacity scenarios (Fig. 2.5).

In Fig. 2.5, BBU stands for building baseband unit, the indoor baseband processing unit. RRU stands for remote radio unit, a remote radio module, which converts digital baseband signal into high-frequency signal and sends the high-frequency signal to the antenna.

Most outdoor 4G mobile network deployments are currently based on macro cells. (In the early days of cellular mobile phone network construction, cells using cellular technology were called "micro cells." Macro cells refer to large areas, where base station transmitting antennas are usually set up above surrounding buildings.) Although macro cells cover large geographic areas, it struggles to provide the dense coverage, low latency, and high bandwidth required for some 5G applications.

To provide the coverage density and high-capacity networks required for 5G, telecom operators are building dense 4G radio access networks (RANs) by deploying micro base stations, especially in some densely populated urban areas. Although the geographic range served by the micro base station is much smaller than that of the wide area base station, it increases the network coverage, capacity, and quality of service (Fig. 2.6).

Micro base stations do not require additional spectrum to increase network capacity, so they are very attractive to operators with low spectrum capacity or spectrum scarcity. In addition, the industry believes that the deployment of micro

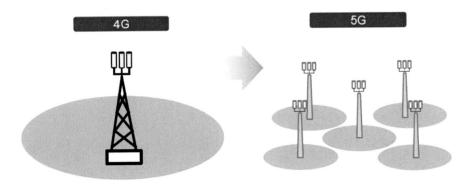

Fig. 2.6 Micro base station

base stations in dense urban areas can improve the quality of existing 4G networks and support the expected high-capacity requirements of 5G and early eMBB services.

Due to the intensive coverage required by the micro base station, there are some telecommunication operators who discuss to install it on public facilities such as bus stops, street lights, and traffic lights, which are usually accompanied by a street cabinet to accommodate the radio equipment, power supply, and field connection of operators.

2.2.3 Massive MIMO

MIMO refers to Multi-Input Multiple-Output, where multiple antennas send and multiple antennas receive. Massive MIMO means that it can be extended to hundreds or even thousands of antennas and supports beamforming to improve data transmission rates, which is critical for efficient power transmission. Massive MIMO not only improves spectrum efficiency but also helps telecom operators meet the challenging capacity requirements of 5G by combining with dense micro base station deployments.

The main reason why 5G has a much higher communication rate than 4G is the large-scale antenna. In 4G and previous 1G, 2G, and 3G eras, antennas were mainly slender, while 5G changes antennas into rectangular or square antenna arrays. The antenna board is located between the base station and the mobile communication terminal and is used to implement high-speed data transmission. Inside it, there are more than 100 antennas arranged at equal intervals, that is, massive MIMO antennas (Fig. 2.7).

In the 4G era, MIMO is already available, but the number of antennas is not large. Instead, it can only be said that it is an initial version of MIMO. MIMO is a technology that achieves high-speed communication through multiple antennas

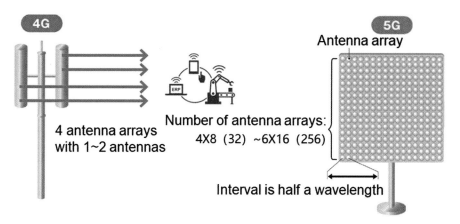

Fig. 2.7 Structure of a massive MIMO antenna

between a base station and a terminal. Multiple antennas transmit the same content of radio waves, and the terminal combines them after receiving. The use of large-scale antennas can not only accurately send complex waves on the one hand but also reduce the attenuation of signal transmission and avoid the impact of obstacles or radio wave interference.

It can be said that in the 5G era, MIMO technology continues to be carried forward and the number of antennas is expanded by more than ten times. Now it has become an enhanced version of massive MIMO antennas.

2.2.4 Device-to-Device

Because the current architecture of wireless mobile broadband communication network is based on fixed infrastructure as base station to communicate with various types of terminals, the communication between terminals needs to be relayed from the base station to the network. Thus, according to Shannon's theorem, due to the limitation of the number of base stations and the wireless access spectrum, there is little room for further capacity improvement of wireless access system. Therefore, it can be predicted that the future development of wireless mobile broadband communication network will encounter significant problems, where wireless access network cannot meet the needs of large-scale increase of wireless mobile data traffic. At present, the flow of information is showing a trend of localization and concentration in hot spots, while the existing wireless access network technology cannot guarantee the quality of service requirements in this application scenario.

5G technology makes D2D (device-to-device) communication possible. D2D technology refers to the direct communication between two peer-to-peer users. In a distributed network composed of D2D communication users, each user can

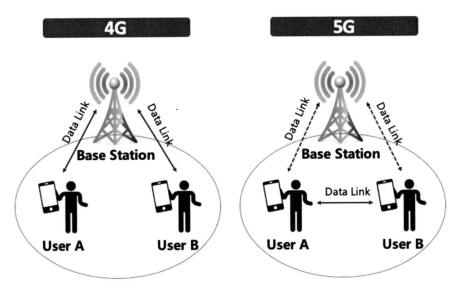

Fig. 2.8 Low-latency D2D communication

send and receive signals and has the function of automatic routing (forwarding messages). Participants in the network share part of their hardware resources, including information processing, storage, and network connectivity. These shared resources provide services and resources to the network and can be accessed directly by other users without passing through intermediate entities. In a D2D communication network, users play both server and client roles, are aware of each other, and self-organize into a virtual or real group (Fig. 2.8).

The advantages of D2D communication technology and its major scenarios are:

2.2.4.1 Local Business: Significant Provision of Spectral Utilization

With the application of this technology, users communicate through D2D to avoid the use of cellular wireless communication, so they do not use band resources. Moreover, the user devices connected by D2D can share the resources of the cellular network and improve the resource utilization.

- Social applications. The most basic scenario for D2D communication technology is social applications based on proximity. With the D2D communication function, data can be transferred between adjacent users such as content sharing and interactive games. Users can find users of interest in adjacent areas through the D2D discovery function.
- Local data transmission. It is achieved by using the proximity and data pass characteristics of D2D, which can save spectral resources and expand mobile communication scenarios. Local advertising services based on proximity can

provide users with information such as discount promotions and movie trailers, to maximize benefits by accurately targeting target users.

- Cellular network traffic offload. With the growing multimedia business with high-definition video and large traffic characteristics, the core layer and spectrum resources of the network are facing enormous challenges. Local multimedia services utilizing the local characteristics of D2D communication can greatly save resources at the core layer of the network and the spectrum. For example, an operator or content provider can set up a server in a hot spot area to store the current popular media business on the server, which provides services to users with business needs in D2D mode, and users can also obtain the required media content from adjacent user terminals that have already obtained the media business, thereby alleviating downstream transmission pressure on cellular networks. In addition, cellular communication between close users can be switched to D2D mode to offload cellular network traffic.

2.2.4.2 Emergency Communications: Ensuring Communications Quickly

D2D communication can solve the problem of communication infrastructure disruption caused by extreme natural disasters, which hinders rescue efforts. In D2D communication mode, wireless communication can still be established between two adjacent mobile devices to provide communication support for disaster relief. In addition, in the blind area covered by wireless communication network, users can connect to the user terminals within the coverage area of the wireless network through one or more hops of D2D communication, with which they can connect to the wireless communication network.

2.2.4.3 Enhancement of IoTs: Expanding the Application of Various Scenarios

According to industry forecasts, there will be about 50 billion cellular access terminals worldwide by 2020, most of which will be machine communication terminals with the characteristics of IoTs. If D2D communication technology is combined with IoTs, it is possible to create a truly interconnected wireless communication network.

2.3 Flexibility in 5G

The three typical application scenarios of 5G have significant differences in network performance requirements. In order to control costs, telecom operators have to choose network slicing/edge computing technologies to achieve the most abundant

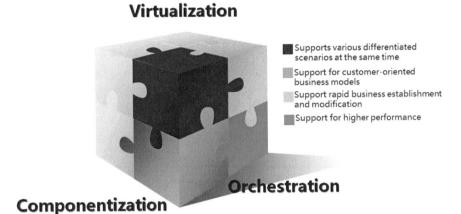

Fig. 2.9 Features of 5G networking

network functions with the least capital investment. A large number of services are completed at the edge of networks (Fig. 2.9).

2.3.1 Edge Computing

5G communication networks are more decentralized, and small-scale or portable data centers need to be deployed at the edge of networks to perform localization requests to meet the ultra-low-latency requirements of URLLC and mMTC. Therefore, edge computing is one of the core technologies of 5G [2]. Edge computing brings data closer to end-user devices, provides extremely low-latency computing capabilities for demanding applications, speeds up the transfer of operable data, reduces transportation costs, and is becoming increasingly important for real-time and latency-sensitive applications (Fig. 2.10).

Generally, edge computing [3] refers to an open platform that integrates core capabilities of networks, computing, storage, and applications on the edge of networks near the source of object or data. It makes full use of the processing capabilities of various devices along the entire path, stores and processes redundant data to reduce network bandwidth occupation, and improves system real-time and availability. Edge computing meets the key needs of industry digitalization in agile connection, real-time business, data optimization, application intelligence, security, and privacy.

Edge computing is to sink the computing and storage capacity of the cloud to the edge of networks and use the distributed computing and storage to directly process and/or solve the specific business needs in the local area, so as to meet the hard requirements of emerging new formats for high bandwidth and low latency. In addition, it

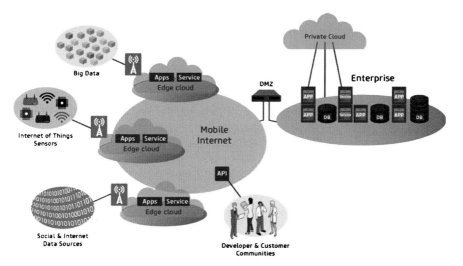

Fig. 2.10 Mobile edge computing. (The source of this figure is from the European Telecommunications Standards Institute.)

- provides information technology service environment and cloud computing capabilities close to mobile users;
- pushes the content distribution to the user side;
- deploys application, service, and content in a highly distributed environment;
- supports low-latency and high-bandwidth service requirements in 5G networks.

In the 4G era, real-time performance, bandwidth, energy consumption, and security are always challenges wanted to be solved by the mobile communications industry. For example, in a cloud computing scenario, after a sensor receives data, it needs to be transmitted to the data center through the network and then analyzed and processed before being fed back to the terminal device by the network. However the data transmission back and forth in this way causes a high delay. For another example in the field of public safety, every HD camera needs 2M bandwidth to transmit video. Such a camera can generate more than 10 GB of data in a day. If all the data is transferred to the data center for analysis and storage, the bandwidth consumption is huge.

Edge computing can perfectly solve many of the above problems. For example, edge computing is deployed on the access network, where data analysis and processing can be completed at the edge of networks. The data does not even need to be uploaded to the cloud, which greatly reduces the data transmission time. It not only reduces the bandwidth pressure but also improves the efficiency and security of the communication network.

Edge computing was actually proposed as early as 2002. In recent years, with the maturity of advanced network technologies such as SDN/NFV, major network standardization organizations have gradually realized that edge computing has greatly improved network functions. In April 2016, ETSI (European Telecommunications Standards Institute) listed it as the key technology of 5G network architecture.

The entire framework of edge computing emphasizes the requirements of "cloud-edge-end" integration. Edge computing collect basic data from the device side to the edge platform, analyze and process the data on the platform, and get instant results back to the device side. The edge manager is responsible for the unified deployment of data, establishing a connection with the cloud, transmitting business-related data to the cloud for deeper analysis, and then optimizing the edge-side algorithms to guide production practices flexibly and efficiently.

Edge computing is a bridge connecting the cloud and end devices. Its geographical location and functional positioning determine its own characteristics and attributes. The sensors collect data and transfer it to edge device, where it performs some real-time processing at the edge layer and transmits it to the core layer for in-depth calculation and analysis. Finally, it returns the analysis results to the edge to optimize and improve the edge intelligence. The two constitute a complete system. Cloud computing is responsible for global, non-real-time, long-period big data processing and analysis, while edge computing processes and analyzes local, real-time, short-period data according to specific needs.

Various applications of 5G in the Internet of Everything scenario are difficult to divide and conquer with terminal computing problems, while edge computing technology undoubtedly carries almost all terminal computing capabilities. In the future, 5G deployment, complemented by the double-wing collaboration of artificial intelligence and the industrial Internet, will lead to a surge in the number of connected devices and an exponential increase in the amount of data on the edge. If all this data is handled by a cloud-based management platform, problems will occur in real time, agility, security, and privacy. While using terminal edge computing, massive data can be processed nearby, and various devices can achieve efficient collaboration.

Under the Internet of Everything, edge computing is an important connection tentacle to achieve interconnection and interoperability. In the application of service industry scenarios, edge computing has many advantages, which can be roughly summarized into two categories:

- The advantages of distributed and low-latency computing, high efficiency, and real-time collaborative can support scenarios such as autonomous driving, smart manufacturing, smart cities, and applications in the AI industry. Taking smart manufacturing as an example, the central intelligent robot is responsible for the overall control in the cloud computing center. The pipeline terminal cooperative robot constantly exchanges and analyzes data to achieve the real-time decision-making and guide other terminal robots upstream and downstream of the pipeline to carry out collaboration, so as to complete the automatic production process with high efficiency and low delay. The ultimate goal of future smart manufacturing is to realize an intelligent factory. As the machine terminal equipment has the ability of self-sensing, it can independently complete the terminal calculation, data processing, data synchronization, and cooperation among terminal equipment. In addition, it can make the optimal decision, complete the complex process operation, and reduce the human intervention as much as possible.

- Edge computing can help alleviate the pressure of data traffic and reduce the data traffic from the device to cloud. According statistics, the cost of using "edge cloud" is only 39% of the cloud cost [4]. Its high-efficiency and energy-saving features can support scenes like AR/VR, UAV, panoramic live broadcast, automatic driving, etc. that generate massive, multidimensional, heterogeneous data in terminals. Taking autonomous driving as an example, the environment inside and outside the car needs to be monitored in real time. The volume of multidimensional dynamic data, such as vehicle networking data, owner behavior data, and road traffic data, generated during the process is huge. For safety reasons, unmanned vehicles need to make real-time decisions for the environment inside and outside the vehicle and guide the vehicle to drive safely or warn the owner of risks in a timely manner to ensure the maximum driving safety.

Edge computing technology is one of the core technologies to solve the diverse network requirements brought by different applications. From the perspective of industrial services and the implementation of actual scenarios, edge computing is driven by 5G to serve vertical industries and specific scenarios. It is inseparable from the close collaboration and complementarity with cloud computing. Specifically, cloud computing focuses on global, non-real-time, long-period big data processing and analysis, while edge computing is suitable for local, real-time, short-period data processing and analysis.

With the further promotion of 5G technology, diversified applications promote the rapid and iterative upgrade of edge computing. Traditional data centers increasingly extend to the edge side, and the computing tasks undertaken by the edge side continue to increase, compared with general-purpose servers. The edge computing server can be personalized and differentiated for 5G and edge computing-specific scenarios, with lower energy consumption, wider temperature adaptability, and more convenient operation and maintenance management. In the 5G era, edge computing technology is introduced in the transmission network architecture. Gateways, servers, and other equipment are deployed in the edge equipment room near the access side to increase computing capabilities. Data such as low-latency services, local data, and low-value data are stored in edge equipment rooms. For processing and transmission, there is no need to return the core network through the transmission network, thereby reducing latency and backhaul pressure and improving user experience.

With the continuous innovation of the underlying network technology, new applications and business models have been introduced. The three major application scenarios for 5G will spawn a large number of different applications in the future, which will place higher requirements on network performance. With the maturity of these applications, higher requirements are placed on network capabilities. New network technologies must be adopted to meet the needs of a large number of

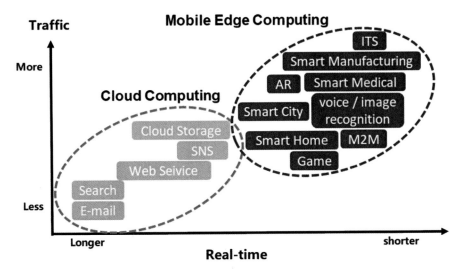

Fig. 2.11 5G and edge computing

different applications. Therefore, it is foreseeable that the development trend of 5G technology will transform the "centralized computing architecture" with cloud computing as the core to the "decentralized computing architecture" with edge computing as the core (Fig. 2.11).

2.3.2 Network Slicing

Network slicing, as a core part of the 5G network architecture, follows the trend of intelligent transformation of operator's edge equipment rooms and is committed to solving the 5G network's hard requirements for low latency, large bandwidth, and massive IoT. Industry-related companies are the key to occupying the 5G development path, grasping new business forms in the future, and exploring more application scenarios.

End-to-end flexibility is one of the defining characteristics of 5G networks. This flexibility is largely due to the introduction of network softwareization, where the core network hardware and software functions are separated. Network softwareization refers to technologies such as network function virtualization (NFV), software-defined networking (SDN), and Cloud-RAN (C-RAN). Based on the network slicing method, the speed of innovation and mobile network transformation is improved.

- NFV replaces network functions on specialized equipment (such as routers, load balancers, and firewalls), and virtualized instances run on commercial off-the-shelf hardware. The cost of network changes and upgrades is also reduced.

- SDN allows real-time dynamic reconfiguration of network elements, enabling 5G networks to be controlled by software rather than hardware. Network resiliency, performance, and quality of service are improved.
- C-RAN is a key disruptive technology that is critical to 5G networks. It is a cloud-based wireless network architecture that uses virtualization technology and a centralized processing unit. It can replace the distributed signal processing unit of a mobile base station and reduce the cost of deploying a dense mobile network based on micro base stations.

Network slicing allows the physical network to be divided into multiple virtual networks (logical segments). These networks can support multiple types of services for different RANs or certain customer groups, thereby greatly reducing network construction costs by using communication channels more efficiently (Fig. 2.12).

On the one hand, different application scenarios require different combinations of network functions to generate different network slices. On the other hand, high-bandwidth and low-latency services need to perform service termination at the edge of the network to generate edge network slices (Fig. 2.13).

5G networks provide solutions suitable for various artificial intelligence and industrial Internet scenarios through network slicing, which also achieve real-time efficiency, low energy consumption, and simplified deployment.

First, the network slicing technology is used to ensure that the network resources are allocated on demand to meet the requirements of the network in different

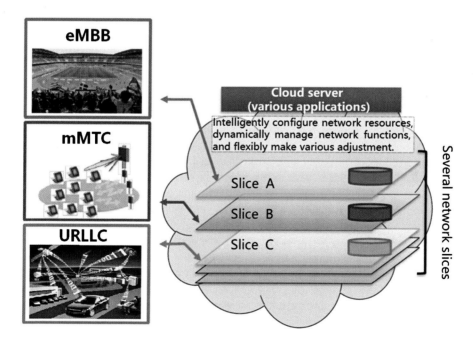

Fig. 2.12 Network slicing enable the three scenarios of 5G

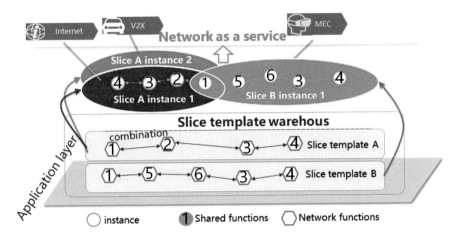

Fig. 2.13 Composition of network slices

manufacturing scenarios. Different applications have different requirements for latency, mobility, network coverage, connection density, and connection cost. More stringent requirements are put forward for the flexible configuration of 5G network, especially for the reasonable and rapid allocation, and redistribution of network resources.

As the most important feature of 5G network, based on the end-to-end network slicing capability combined with a variety of new technologies, the required network resources can be flexibly and dynamically allocated and released to different needs in the whole network. According to the blueprint and input parameters provided by service management, network slicing is created to provide specific network features. For instance, extremely low latency, high reliability, and huge bandwidth can meet the requirements of different application scenarios on the network. In the smart factory prototype, in order to meet the key transaction processing requirements in the factory, key transaction slices were created to provide a low-latency and highly reliable network.

In the process of creating a network slice, resources in the infrastructure need to be scheduled, including access, transmission, and cloud computing resources. Each infrastructure resource also has its own management function. According to the different needs of customers, networking slicing provides ether shared or isolated infrastructure resources. Due to the independence of various resources, network slice management also performs collaborative among different resources.

In addition to key transaction slices, 5G will also create mobile broadband slices and large connection slices. Different slices share the same infrastructure under the scheduling of the network slice management system. They are not interfering with each other and maintaining the independence of their respective services.

Furthermore, 5G can optimize network connections and take local traffic offload to meet low-latency requirements. The optimization of each slice for business needs is reflected not only in different network functions and features but also in flexible deployment schemes. The deployment of network function modules inside the slice is very flexible and can be deployed in multiple distributed data centers, respectively, according to business needs.

Finally, it adopts distributed cloud computing technology to deploy industrial applications and key network functions based on NFV technology in a local or centralized data center in a flexible manner. The high-bandwidth and low-latency characteristics of 5G networks have greatly improved intelligent processing capabilities by migrating to the cloud and paved the way for improved intelligence.

References

1. Ofcom, "Update on 5G spectrum in the UK." https://www.ofcom.org.uk/__data/assets/pdf_file/0021/97023/5G-update-08022017.pdf. Accessed: 2020-03-31.
2. L. Gao, T. H. Luan, B. Liu, W. Zhou, and S. Yu, *Fog Computing and Its Applications in 5G*, pp. 571–593. Cham: Springer International Publishing, 2017.
3. T. H. Luan, L. Gao, Z. Li, Y. Xiang, G. We, and L. Sun, "A view of fog computing from networking perspective," *ArXiv*, vol. abs/1602.01509, 2016.
4. P. R. K. S. Stephen Belanger, Ashish Nadkarni, "Market analysis perspective: Worldwide core and edge computing platforms, 2018," *IDC*, 2018.

Chapter 3
Development of Industry 4.0

This chapter dissects the practical path of industrial 4.0: its core, vision, and key technologies of intellectualization, as well as the key production elements and tools of future industrial manufacturing enterprises. For decades, with the informatization, digitization, and networking of various products, the structure of manufacturing industry itself has become more complex, finer, and automated. The data flow inside production lines and equipment, as well as the data of management work, has increased dramatically. In the past, the automation system was unable to meet the higher demands of the manufacturing industry in terms of information processing capacity, efficiency, and scale. The applications of next-generation information technologies such as 5G, cloud computing, IoTs, and AI have solved this problem to a certain extent.

Today, the world is witnessing a new wave of industrial revolution and innovation marked by "intelligence." The core of the change is the application of next-generation information technologies. When the new form began to be injected into the manufacturing field, a new challenge also began, that is, the challenge of turning manufacturing from mechanization, electrification, and digitization to networking, data, and intelligent manufacturing.

3.1 The Challenges of Traditional Industrialization

In the twentieth century, the large-scale production model occupied the dominant position in the global manufacturing industry. It once greatly promoted the rapid development of the global economy and brought the entire society to a whole new stage. However, with the development of the world economy and the increasingly fierce market competition, consumers' consumption views and values are becoming more and more diverse and personalized. With this, the uncertainty of market

© Springer Nature Singapore Pte Ltd. 2020
X. Wang, L. Gao, *When 5G Meets Industry 4.0*,
https://doi.org/10.1007/978-981-15-6732-2_3

demand has become more and more obvious, and large-scale production methods can no longer adapt to this rapidly changing market environment.

It can be said that the past 30 years have been the fastest period of globalization. Developed countries have transferred a large number of labor-intensive industries to developing countries with relatively low labor costs and raw material costs through industrial transfer. And the rise in labor costs and raw material costs also pose tremendous pressure on the future manufacturing industry objectively.

In addition, affected by the external environment, such as the relative shortage of resources, increased environmental pressure, and intensified excess production capacity, the traditional industrial economy driven by energy conversion tools will be difficult to maintain. Environmental pollution, ecological damage, and the scarcity of resources and energy have become severe challenges facing human society. Solving these global social problems and achieving sustainable development have become the consensus of humankind (Fig. 3.1).

In other words, under the pressure of cost, resource, environmental constraints and the market, traditional industrialization development model has lost its competitiveness.

As a major manufacturing country, Germany started to implement the national strategy of "Industry 4.0" in April 2013. It hopes to apply Internet technology in all links in the future manufacturing industry, visualize the connection between digital information and the real society, and fully integrate management processes. As a result, smart factories are realized and smart products are produced. "Industry 4.0" is also considered to be the fourth industrial revolution designed to support the development and innovation of a new generation of revolutionary technologies in the industrial field.

Is the ongoing "Industry 4.0" an opportunity or a challenge for us? In the market of manufacturing sector, companies that adopt new business models are bound to emerge. Traditional manufacturing may still exist in the market, but in order to cope with new competitors, their managers will definitely change their organizational

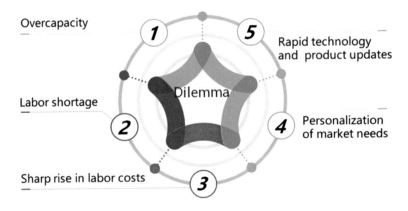

Fig. 3.1 Pressure on manufacturing

structure, management processes, and business functions during the industrial revolution. Smartphones and wearable devices are not only successful because they are new, but more importantly, the consumer culture and social transformation that followed.

In the first industrial revolution, the invention of steam engine was mechanized. The second industrial revolution was the invention of electricity, enabling electrification. Beginning in the 1970s, with the development of information technology, the application of software systems such as computer service systems and ERP (enterprise resource planning) in the field of manufacturing had brought digitalization and automation to manufacturing industry. It can be said that the first three industrial revolutions have continuously evolved the production mode of manufacturing. "Industry 4.0" has changed the paradigm of manufacturing based on the third industrial revolution.

The manufacturing industry in the past was only one link, but as the Internet penetrated further into the manufacturing link, network collaborative manufacturing has begun to emerge. The manufacturing model will change drastically with it, which will break the life cycle of traditional industrial production. From the purchase of raw materials to the design, research and development, production and manufacturing, marketing, after-sales service, and other links of products form a closed loop, which completely change the production mode of manufacturing industry to only one link. In the closed loop of network collaborative manufacturing, the roles of users, designers, suppliers, distributors, etc., will change. Along with it, the traditional value chain will inevitably break and reconstruct.

"Industry 4.0" represents a new round of industrial revolution. Intelligent manufacturing is the development of a more efficient and refined future manufacturing. Information technology has made the manufacturing industry from digital to networked and intelligent. At the same time, the boundaries of traditional industrial fields have become increasingly blurred, and industrial and non-industrial will gradually become difficult to distinguish. The focus of the manufacturing link is not on the manufacturing process itself, but on the individual needs of users, product design methods, resource integration channels, and network collaborative production. Therefore, some information technology companies, telecommunications operators, and Internet companies will be closely connected with traditional manufacturing companies, and it is very likely that they will become leaders in traditional manufacturing companies.

3.2 Traditional Industrialization and AI

AI is a new technological science that studies and develops the theories, methods, technologies, and application systems used to simulate and extend human intelligence. Since the 1950s, AI has been coming into everyone's view. There are also many different understandings of AI in academia and industry. The diversified development of science, technology, and commerce has led to different understandings of the definition, driving force and manifestations of AI. Generally speaking, AI can be divided into human-like behavior (simulated behavior results), human-like thinking (simulated brain operation), and generalized (no longer limited to simulated human) intelligence.

2016 is a "milestone" year for AI. At the beginning of this year, AlphaGo won the professional 9-dan by Lee Sedol, bringing the AI technology, which has risen again in the past 10 years, to the stage and into the public's view. In the past few years, giants of science and technology have established AI laboratories successively, investing more and more resources to seize the AI market. Even more, some enterprises have been transformed into AI companies as a whole, to step up planning for the future layout of AI. Many countries regard AI as the strategic leading factor in the future, formulate strategic development plans, promote it from the national level as a whole, and welcome the coming era of AI.

From human-machine game to smart home, from simultaneous interpretation to face recognition, AI that slightly "science fiction" decades ago has now really integrated into our lives.

AI has arrived and it's right next to us, almost everywhere. With the success of in-depth learning in the fields of computer vision, speech recognition, and natural language processing, many applications relying on AI technology have matured and begun to penetrate into all aspects of our lives, both in consumers and enterprises. Smart recommendation systems, as small as smartphones we use and as large as smart cities, smart transportation, smart security systems, and smart financial investment systems, all rely on AI technology based on machine learning algorithms. AI algorithms exist on people's mobile phones and personal computers, on servers of government agencies, enterprises and public welfare organizations, and on common or private cloud computing platforms. Although we may not be able to truly perceive the existence of AI algorithms, they have been highly penetrated into our lives. It is because of these algorithms that AI can be used more widely.

AI represents one of the main directions of scientific and technological innovation in the future. Especially under the joint drive of new theories and technologies such as mobile internet, big data, supercomputing, IoTs, brain science, and strong demand of economy and society, this has triggered the third wave of AI development today.

Since the concept of AI was proposed in 1956, it has experienced three development climaxes. The first phase of AI takes neuron model and Turing test as its representative but lacks theoretical and technical support, so that it fell to a low ebb in the 1980s. With the development of expert system and machine learning in the 1990s, artificial labor can only start to industrialize, ushering in the second climax. But the prospects are not long. Because the AI infrastructure is not perfect, the technology is not mature enough, AI has lost its popularity again in around 2000. Until 2010, with the development of new generation AI technologies such as deep learning and federated learning, it has reached its third climax and entered everyone's view (Fig. 3.2).

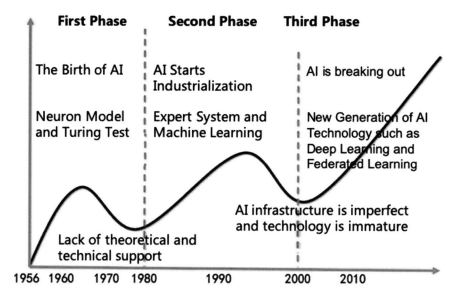

Fig. 3.2 A brief history of AI

Throughout the history of science and technology in the world, many technologies developed slowly in the first stage and could not be upgraded for a long time. Even worse, they often deviate from straight-line growth expectations. Until the second stage, there was a sudden rapid development at a certain point in time, and it suddenly caught up with the straight-line growth level. The third phase, when everything is ready, will develop rapidly, reaching infinitely vertical growth – as is the case with the new generation of AI. The drivers of AI are mainly algorithm/technology, data/computing, scenarios, and disruptive business models driven. With the upgrading of the algorithm, the explosive growth of big data and the landing of the application scenario, it can be predicted that AI will show exponential explosive growth in the next few years.

This rise in AI is not just about laboratory research. Research and commercialization of theory and key commonality technologies have been pushed forward simultaneously, which has resulted in the emergence of more cases of product-oriented, solution, and service-oriented applications in AI, making the public truly feel its existence. Especially in the areas of image analysis, speech recognition, satellite navigation, and natural language processing based on in-depth learning algorithm applications are rapidly industrializing, and the track of industrial competition is also opened.

There are three main reasons why the new generation of AI is maturing in this era, which are the rapid development of the new generation of information technology, the explosion of new social needs, and the change of basic goals (Fig. 3.3).

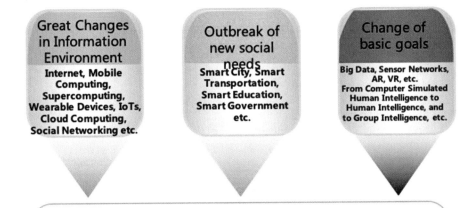

Fig. 3.3 Drivers of the new generation of AI

- From the perspective of the new generation of information technology, 5G, big data, mobile computing, supercomputing, wearable devices, IoTs, cloud computing, social network, and search engines driver AI upgrade.
- From the explosion of new social needs, smart cities, smart medical, smart transportation, smart education, smart environmental protection, smart government, and other consumer upgrades need AI.
- From the change of basic goals, big data, multimedia, sensor network, AR, and VR. The next generation of AI will range from computer-simulated human intelligence to human intelligence and to group intelligence.

The new generation of AI has also brought the industry into the AI era. For example, with the increasing degree of intelligent manufacturing, the development of intelligent robot technology has entered an explosive phase, and the scope of intelligent application has been greatly expanded. In addition, new technologies and new industries such as speech recognition, face detection, unmanned driving, intelligent robots, smart hardware and software, smart terminals, virtual reality, and augmented reality are also developing rapidly.

It can be said that the new generation of AI is triggering a new round of industrial revolution represented by industrial 4.0 (Fig. 3.4). Strongly restructuring economic activities such as R&D, production, marketing, and consumption of industries will have a significant impact on productivity, production relations, economic base and superstructure, as well as on the political, economic, social, cultural, and life of the future.

Fig. 3.4 Industry entering AI era

3.3 Algorithm: The Core of the Industry 4.0

At present, information systems have been applied to a certain degree in the company's design and production processes, for example, the usage of computer-aided design (CAD) in research and development aspects and the usage of ERP in business management. However, in the process of solving the design, manufacturing, management, and other aspects of information collaboration, the application system solutions still need to improve.

The smart factory refers to simulate, evaluate, and optimize the entire production process in the computer virtual environment, which based on the products and related factory data of full life cycle basis. It is further extended to the new production organization throughout the product and factory life cycle.

With the development of artificial intelligence, cognitive computing, and machine learning, the system has been able to interpret, adjust, and learn the data collected by interconnected devices. With the ability of development and adaptation, combined with powerful data processing and storage capability, manufacturers can not only automate tasks but also deal with highly complex interconnected processes. With increasingly global supply chains and discrete manufacturing process and production, production processes often involve multiple devices and different parts supplier. At the same time, supply chains become more complex due to the demand of regional, local, and individual needs. Industry 4.0 is actually based on the cyber physical systems (CPS) to realize the intelligent factory, and the final realization is the change of manufacturing mode.

Industry 4.0 evolved from embedded systems to CPS and form smart factories. As the representative of the fourth industrial revolution in the future, smart factories continue to develop toward the seamless Internet (IoTs, data network, and service internet), which realizes objects, data, and services. In the future, smart factory needs to get through the information barriers of different layers such as equipment, data collection, enterprise information system, and cloud platform, to realize the vertical interconnection from the workshop to decision-making layer (Fig. 3.5, where IT refers to information technology and OT refers to operational technology).

The Internet of Things and the Internet of Services are respectively located at the bottom and top of the three-layer information technology infrastructure of the smart factory. ERP, SCM, CRM, etc., which related to production planning, logistics, energy consumption and management, and PLM, which related to product design and technology, are at the top level and closely connected to the service Internet. In the middle layer, the CPS is used to implement functions related to production equipment and production line control. From the supply of intelligent materials to the output of intelligent products, it runs through the entire product life cycle management. At the bottom level, control, execution, and sensing are implemented through the Internet of Things technology to achieve intelligent production. For example, today's factory material distribution system is no longer as simple as automatically circular pick-up according to the predetermined route. The logistics link is not only developing toward the real unmanned direction but

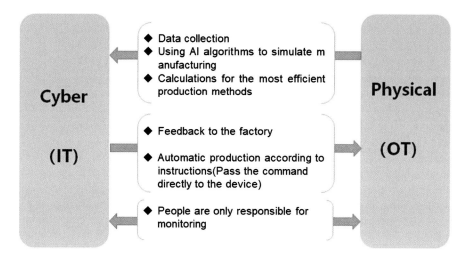

Fig. 3.5 Cyber physical systems

also identifying the needs through an interconnected system, communicating the instructions to an autonomously controlled transportation system, and realizing the real-time response. These systems transfer information and data to each other and to the networked workbench and warehouse, so as to respond to the change of supply and demand dynamically.

The highly integrated smart factory can greatly improve the production efficiency of the enterprise, effectively organize the resources of all parties, encourage the production enthusiasm of employees in different chains, change the enterprise from different individuals into a team with super cohesion, and make fundamental changes in personnel organization and management, task allocation, work coordination, information exchange, design data, and resource sharing. Industry 4.0 itself advocates the integration of production equipment, sensors, embedded systems, production management systems into an intelligent network through CPS, so that equipment and services can be interconnected to achieve horizontal, vertical, and end-to-end integration.

Horizontal integration refers to a mechanism of resource information sharing and resource integration between enterprises of network collaborative manufacturing through value chain and information network, which ensures seamless cooperation among enterprises and provides real-time products and services. Horizontal integration is mainly reflected in the network cooperation, which mainly refers to the integration based on enterprise business management system, such as inter enterprise industrial chain and enterprise group, to generate new value chain and innovation of business models.

Vertical integration refers to the realization of decentralized production based on a networked manufacturing system in a smart factory, replacing the traditional centrally-controlled production process. Vertical integration is mainly reflected in the scientific management in the factory, from focusing on the design and manufacturing process of products to the integration process of the entire life cycle of products and establishing an effective vertical production system.

End-to-end integration refers to the integration of engineered information systems throughout the entire value chain to ensure the implementation of large-scale personalized customization. The end-to-end integration is oriented by the value chain, realizes the end-to-end production process, and enables the effective integration of the information world and the physical world. End-to-end integration is an examination of smart manufacturing from the perspective of process flow, which is mainly reflected in parallel manufacturing. Through the integration platform, the unit technology products will not form the enterprise's integration platform system but also develop the factory's comprehensive capability platform.

The three integrations of the smart factory, from information sharing as the focus of integration for many years, to the stage of process integration, continue to move forward to the integration stage of intelligent development. It promotes the vertical, horizontal, and end-to-end integration at the existing high-end level, including all-round integration within the enterprise, between the enterprise and the network cooperative enterprises, and between the enterprise and customers. Some case studies are:

Case Study 1, Bosch Rexroth: Mass Customization Through the Internet Bosch Rexroth's Humboldt base, with a history of 50 years and about 700 employees, is one of the first industrial 4.0 demonstration pilots in Germany.

Bosch Rexroth adopts various innovative technologies, including self-guided products, independent work units, automatic identification of employees and products, real-time quality detection, etc., to improve production efficiency and product output by cooperating with RFID identification technologies sorting based on electronic tags (RFID, radio frequency identification) and can meet more varieties of customized production on the assembly line.

By using these technologies in the production line and improving the flexibility, batch customization is economically possible. Industrial manufacturing enterprises and other customers are currently customizing Bosch's hydraulic motion control unit in their own production lines to ensure subsequent production. Bosch Rexroth, through closer integration of product regional assembly technology and mobile controls, has laid the foundation for the transformation from a parts supplier to a flexible module supplier, so that it can not only vertically integrate the supply chain but also provide opportunities for future development of value-added services such as remote diagnosis, predictive maintenance, and pay-per-use.

Bosch Rexroth also provides workers with a variety of tools, including mobile apps, to access all kinds of information when they need it. Real-time quality inspection is also increasingly popular. This technology enables workers to recognize and correct errors in time, and the accuracy of the production process will automatically be checked repeatedly.

Case Study 2, Tesla: "Super Factory" Tesla is a typical representative of smart factories and smart products. Tesla super factory is located in California, the United

States. It was originally a joint venture factory of General Motors (GM) and Toyota. In 2010, it was acquired by Tesla, and a large number of industrial robots were used. It then became the most advanced electric vehicle production factory in the world. All processes from the beginning of raw material processing were carried out in the factory. There are more than 160 robots in the super factory, which belong to stamping production line, car body center, paint center, and assembly center. The multitasking robot in the car body center is the most advanced and frequently used robot, which can perform many different and difficult tasks, including stamping, welding, riveting, gluing, etc.

Tesla's definition of the product is a comprehensive large-scale mobile intelligent terminal with unique human-computer interaction mode, including hardware, software, content, and services. Compared with traditional vehicles, the parts of electric vehicles are greatly reduced. Tesla Model S and Model X have 60% common parts, and the production efficiency is greatly improved. Robots and other digital technologies also make workers easier, safer, and more efficient. Collaborative robots do more than just perform pre-programmed tasks, instead also the workers can "train" these robots in an interactive way. They don't need to spend a lot of time programming, just repeat their actions.

Case Study 3, Komatsu, Japan: KOM-MICS Enables Predictive Maintenance
Komatsu's smart factory system is named KOM-MICS (Komatsu Manufacturing Innovation Cloud System). All production equipment enables network coverage to achieve production information visualization. Komatsu first applied KOM-MICS to its welding robot. In the future, visualization of equipment operation will be realized on all production equipment. Not only KOM-MIC is applied in Komatsu factory, but also Komatsu suppliers will apply this system in the future.

KOMTRA, Komatsu's product remote monitoring system, can learn about the operation of Komatsu products and inform users in time when the products need repair and maintenance and improve the operation efficiency. The KOMTRAX system helps Komatsu expand the soft service market and enhance the added value of the original hardware products.

The intelligent device described in Industry 4.0 mainly refers to the robots (workstations) engaged in operations can access all relevant information in real time through the network, and according to the information content, autonomously switch production methods and replace production materials to adjust the production operation to the most matching mode. The dynamic configuration production method can realize different design, component composition, product order, production planning, manufacturing, logistics, and distribution for each customer and product, eliminating wasteful links in the entire chain. Different from the traditional production method, the dynamic configuration production method can change the original design scheme at any time before or during the production process.

For example, the current automobile production is mainly based on the production line and mode are designed in advance. Although there are some mixed flow production methods, the production process must be carried out in the production line composed of many machines, so the diversification of product design will not be realized. MES (manufacturing execution system), which manages these production lines, should have brought more flexibility to the production lines. However, restricted by the hardware of many machines that make up the production lines, MES can not play more functions, and its role is extremely limited.

At the same time, workers operating on different production lines are distributed in various workshops, and they will not master the entire production process, so they can only play a role in a certain fixed job. As a result, it is difficult to meet the needs of customers in real time.

In the smart factory depicted in Industry 4.0, the concept of fixed production lines disappears, and the "smart device" replaced it with a modular production method that can be dynamically and flexibly restructured.

For example, the production module can be regarded as a "CPS," and the car under assembly can shuttle between the production modules automatically to accept the required assembly work. Among them, if there is a bottleneck in the supply of production and parts, the production resources or parts of other models can be dispatched in time to continue production. That is, for each vehicle model, the appropriate production module is selected for dynamic assembly operations. Under this kind of dynamic configuration production mode, the original integrated management function of MES can be brought into play, and the whole production process of design, assembly, and test can be managed dynamically, which not only ensures the operation efficiency of production equipment but also enables the diversification of production types.

According to the calculation of SAP company, the production efficiency of this modular is expected to increase by 20% [1]. The characteristics and value of new modular production are mainly reflected in four aspects:

The first is independence Each workstation is a separate module and the order restriction of traditional production lines no longer exists. When needed, modules can be added and exited at any time without mutual influence.

The second is variable Each product can have its own virtual variable processing flow sequence. When leaving each workstation, it can make the optimal decision on the destination of the next workstation.

The third is intelligence All products and materials are automatically transported in the workshop through AGV (automated guided vehicles) and will be delivered only when necessary, so as to minimize the quantity under production and improve efficiency.

The fourth is flexibility Modularity improves the scalability and adaptability of the production system and has a higher adaptability to the shape and size of the product, which can be easily adjusted according to the needs.

This modular production is "algorithmic production," which means a flexible production method that handles complexity and dynamic configuration in real time (Fig. 3.6).

Intelligent devices can also be shared among machines, networks, individuals, or enterprises to further promote intelligent collaboration, so that many related

Fig. 3.6 Intelligent
production based on AI
algorithms

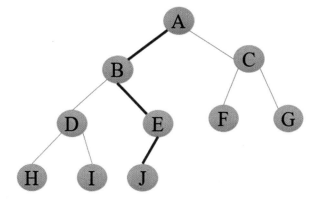

enterprises can participate in the process of asset maintenance, management, and optimization and also ensure that local and far-reaching people with machine expertise are integrated at the right time. Each intelligent device will produce a large number of data that can be transmitted to remote machines and users through the industrial Internet. The data generated by intelligent devices also includes the external data that can optimize the operation or maintenance of machines and units. As time goes on, these data enables the machine to get inspiration from its historical data and operate more intelligently through the control system.

For example, managers can learn how long a particular component has been running under certain conditions. The analysis tool can compare these data with the operation history data of similar components in other factories and provide scientific prediction for the possibility and time of component failure. Thus, the operation data and prediction analysis can be combined to effectively avoid operation interruption and reduce maintenance costs. Once a large number of intelligent data is collected by intelligent devices, intelligent decision-making with business value can be mined through intelligent system. The combination of equipment and data, together with network collaboration and real-time update, will bring great benefits to many industries.

3.4   Break the Impossible Trinity of Manufacturing Industry: The Vision of Industry 4.0

Since the last century, the large-scale production model has dominated the global manufacturing industry. It has greatly promoted the rapid development of the global economy and brought the entire society to a whole new stage. However, with the development of the world economy and the increasingly fierce market competition, consumers' consumption concepts and values are becoming more and more diverse and personalized. Subsequently, the uncertainty of market demand is more and more obvious, and the large-scale production mode has been unable to adapt to this rapidly changing market environment.

In particular, in the late 1970s, automatic control systems began to be used in production. Since then, many factories have been exploring how to improve efficiency, quality, and flexibility of production. Some factories put forward mechatronics and management control integration from the perspective of mechanical manufacturing. Mechatronics realizes the assembly line process, operates in sequence, provides technical support for mass production, and improves production efficiency. Management and control integration based on central control can achieve centralized management, to a certain extent, which saves the cost of production and manufacturing, and improves production quality. However, neither of them can solve the problem of manufacturing flexibility.

In fact, the concept of flexible manufacturing system (FMS) was put forward as early as the 1960s in Britain. FMS mainly refers to a manufacturing system which adapts to the variety of products in an agile way based on the principle of cost-effectiveness and automation technology. According to the data, the flexible manufacturing system is controlled by computer, which is composed of several semi-independent workstations and a material transmission system. Based on the modular and distributed manufacturing units, it can efficiently manufacture many kinds of small batch products through flexible processing, transportation, and storage and can be used for different production tasks at the same time. This kind of manufacturing system with distributed and self-discipline management unit, where each unit has certain decision-making autonomy and has its own command system for planning, scheduling, and material management. It forms a local closed-loop and adapts to the needs of frequent changes in production varieties, which makes the equipment and the whole production line has considerable flexibility. FMS is a kind of reconfigurable advanced manufacturing system which is mainly based on information and has nothing to do with batch. It realizes the transition from "rigidity" to "flexibility" of machining system.

Nowadays, with the rapid development of information technology, computer, and communication technology, people's demand for products changes, making flexibility further become the biggest challenge in the field of production and manufacturing. Specifically, due to the rapid development of technology, and frequent product updates, product life cycle is becoming shorter and shorter. For manufacturing factories, it is necessary to consider not only the ability of quick response to product upgrading, but also the reduction of product batch due to the shortening of life cycle. It however comes the problem of cost increase and price pressure.

The new development mode of manufacturing industry needs to convert challenge of flexible production into an opportunity. It needs to use automation technology to solve the problem of flexible production by integrating with the rapid development of Internet, IoTs, and other information technologies. It requires not only to meet the needs of personalization and customization but also to obtain the advantages of low cost, high efficiency, and fast delivery of large-scale production.

In the past, the manufacturing industry was just a link, and the cooperation between upstream and downstream has always been based on a fixed and simple chain. With the further penetration of the Internet into the manufacturing sector, network collaborative manufacturing has begun to appear. The mode of manufacturing industry is changing greatly. It breaks the life cycle of traditional industrial production, which is a closed loop from the purchase of raw materials to the design, R&D, production and manufacturing, marketing, after-sales service, and other links of products. It will completely change the production mode of manufacturing industry, which has only one link, and solve the impossible trinity of manufacturing industry through intelligent configuration (Fig. 3.7).

Automation is just a simple control. While intellectualization is based on the control, it collects mass production data through IoT sensors and analyzes and mines big data through information management system, so as to make correct decisions. These decisions are added to the automation equipment by "intelligence," so as to improve production flexibility and resource utilization, enhance the close relationship between customers and business partners, and enhance the commercial value of industrial production.

In the future, the roles of users, designers, suppliers, and distributors will change in the closed-loop of network collaborative manufacturing. Along with it, the traditional value chain will inevitably break and reconstruct.

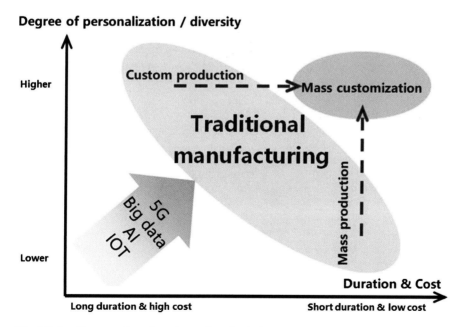

Fig. 3.7 Intelligent configuration of manufacturing industry

3.5   ICT Technology: The Key to Industry 4.0

"Industry 4.0" is actually the realization of smart factories based on CPS and ultimately achieves a paradigm change in manufacturing.

3.5.1   What Is the CPS?

The CPS concept was first proposed by the US National Science Foundation (NSF) in 2006. It is considered to be the third wave of information technology in the world after computers and the Internet. Its core is the integration of 3C (computing, communication, and control) (Fig. 3.8).

CSP is a fusion technology that includes computing, communication, and control (sensors, actuators, etc.). In a broad sense, it is a controllable, trusted, and scalable networked physical device system that deeply integrates computing, communication, and control capabilities on the basis of environmental awareness. It implemented deep integration and real time interaction through a feedback loop in which computing processes and physical processes interact with each other to add or expand new functions, and monitor or control a physical entity in a safe, reliable,

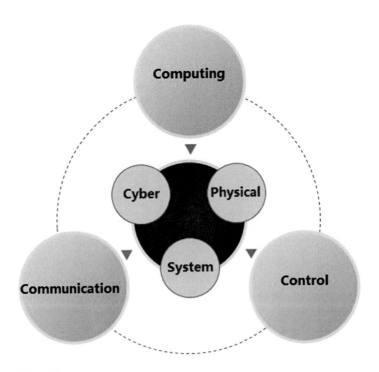

Fig. 3.8 3C in CPS

efficient, and real-time manner. The ultimate goal of CPS is to achieve the complete integration of the information world and the physical world, build a controllable, trusted, scalable, safe and efficient CPS network, and ultimately fundamentally change the way humans build engineering physical systems.

In May 2005, the US Congress requested the National Academy of Sciences to evaluate the technological competitiveness of the United States and put forward proposals to maintain and improve this competitiveness. Five months later, a report based on this study, "Standing on top of the storm," appeared. Based on this, the American Competitiveness Initiative was released in February 2006, and CPS was listed as an important research project. By July 2007, the President's Council of Advisors on Science and Technology (PCAST) listed eight key information technologies in a report entitled "Leadership Under Challenge: Information Technology R&D in a Competitive World," of which CPS ranked first. The rest are software, data, data store and stream flow, networking, high-end computing, cyber security and information assurance, human-computer interaction, network information technology (NIT), and social science.

CPS connects the information world with the physical world, making intelligent objects communicate and interact with each other and creating a real network world. Embedded systems in production equipment and IoT sensors on the production line are one of the elements that compose CPS. These technologies are called "physical technologies." However, CPS embodies a further evolution of the current embedded system and IoTs. It combines with the Internet or the data and services that can be collected on the Internet to realize a more extensive new physical space based on innovative applications or processes and to weaken the boundary between the physical world and the information world. That is, CPS realizes "industry 4.0" by providing the basic part of building IoTs and integrating with "service Internet." Making the physical technology in the traditional manufacturing industry just like the Internet changes the relationship between personal communication and interaction, which will bring new and fundamental changes to the interaction between us and the physical real world.

Once, the interaction between the embedded system based on high-performance software and the professional user interface integrated in the digital network, it will create a new world of system functionality. A simple example is smartphone, where it includes many applications and services that go far beyond the device's own calling function. As new epoch-making application and service providers will continue to emerge and gradually form a new value chain, CPS will also bring paradigmatic changes to existing business and market models. The automotive industry, the energy economy, and various industrial sectors, including production technologies such as "Industry 4.0," will undergo dramatic changes in response to these new value chains.

3.5.2 Industry 4.0 Improves CPS from 3C to 6C

To put it simply, CPS is an embedded system plus a network control function, of which the network function is mainly for achieving control purposes. Further management and control are realized by using the wireless connection and sensing functions of the IoTs and sensors.

In the "Industry 3.0" era, automation systems are closed cycle, and many embedded devices do not reserve external interfaces. "Industry 4.0" expands embedded systems through network functions on the basis of automation, making CPS, which integrates computing, communication, and control capabilities, become the core of smart factories.

In the "Industry 4.0" era, the concept of 3C integration of CPS has been further enriched. On the basis of computing, communication and control, content (Semantic Analysis), community (Collaborative Cooperation), and customization are added (Fig. 3.9). It can also be said that from 3C to 6C reflects the changes in manufacturing thinking and modellings. Under the condition of 6C, the smart factory can realize the intelligent production of the entire industrial chain, achieve the self-regulation of production, and produce intelligent products through the network collaborative manufacturing.

At the same time, the smart factory under 6C conditions can realize visual production and predictive manufacturing management. In traditional manufacturing process, there are many factors that cannot be quantified, including the degradation of equipment performance during processing, sudden failures of parts, rework of defective products, etc. Through visualization, real-time monitoring of production data and control of those uncertain factors, smart factory managers can objectively evaluate the use of manufacturing and equipment, and achieve predictive manufacturing through management, which plays a role in reducing costs, and improving operation efficiency and product quality.

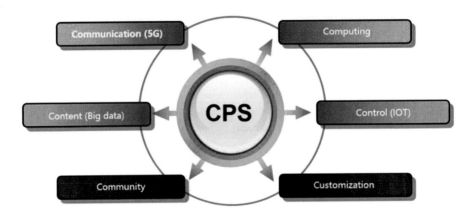

Fig. 3.9 CPS in the " Industry 4.0 " era is the convergence of 6C

3.5.3 CPS Realizes the Integration of Production Process and Information System

The so-called manufacturing informatization currently emphasizes industrial software such as CAD (computer-aided design), CAM (computer-aided manufacturing), PPS (production planning system), and PLM (product life cycle management). It is mainly used in the top-down centralized control system.

"Industry 4.0" realizes the integration of production process and information system through CPS, which reflects the paradigm shift of production mode from "centralized" to "decentralized." All these become possible due to the technological progress that subverted the traditional production process theory. At the same time, distributed intelligent utilization, which represents the interaction between virtual world and physical world in the manufacturing process, plays an important role in the construction of intelligent object network. In the future, the production equipment is no longer just "processing" products; instead, the intelligent production equipment is closely connected through the form of network, and the product communicates to the production equipment through network communication on how to take the correct operation. It is more dynamic and flexible, so as to explore more optimization possibilities and improve production efficiency.

In the future of intelligent manufacturing, CPS is of great significance to cover many industrial sectors and applications such as automation, production technology, automotive, mechanical engineering, energy, transportation, and telemedicine. CPS can not only reduce the actual cost, improve the efficiency of energy, time, etc., but also reduce the level of CO_2 emissions, and play a significant role in protecting the environment. Many applications enabled by CPS will generate new added value chains and business models.

Because the existence of CPS, the processing of production systems and smart factories will have a very high level of real-time performance. At the same time, they will also have advantages in resource and cost savings. Therefore, features such as flexibility, self-adaptation, machine learning capabilities, and even risk management are indispensable elements. The equipment of the smart factory will realize advanced automation, which is mainly achieved by the flexible network of CPS production system based on automatic observation of the production process. Through a flexible production system that can respond in real time, the production process can be completely optimized. At the same time, production advantages are not only reflected in one-time under specific production conditions but also the optimized world-class network formed by multiple factories and production units.

3.6   Big Data: The Main Production Element of Industry 4.0

3.6.1   An Era of Data Explosion

In recent years, with the rapid development of information technology and communication technology such as the Internet, IoTs, cloud computing, and mobile communication, the explosion of data volume has become a serious challenge and valuable opportunity faced by many industries. Jack Ma, the founder of Alibaba Group in China, spared no effort to promote his views, "People are moving from the IT era to the DT (Data Technology) era," on various occasions, and the information society has entered the Big Data era. The emergence of big data has changed not only people's life and work mode but also the operation mode of enterprises. Jack Ma believes that the IT era is dominated by self-control and self-management, while the DT era is dominated by technology that serves the public and stimulates productivity. There seems to be a technical difference between the two, but it is actually a difference in the ideological level.

Big data means that the amount of data involved is too large to be fully collected, managed, processed and analyzed within a reasonable time using the current computer software tools. Mainly, there are a lot of data born on the Internet every day, including a lot of data generated with the popularity of social networks and e-commerce transactions, as well as data such as user location and life information collected by the mobile Internet, smartphones, tablets, and a variety of IoTs sensors all over the globe. From the data attributes point of view, large data is a data collection consisting of a large number, complex structure, and a large number of types. Large data are typically characterized by 4 V (Fig. 3.10).

- Volume: The volume of data is huge. The size of large data is still a changing indicator and the size of a single dataset can range from dozens of TBs to a few PBs (storing 1PB data requires 20,000 PCs with a 50 GB hard drive). Moreover, data can be generated from all kinds of unexpected sources. According to IDC, a US market research company, global data volumes will grow 50-fold by 2020.

Fig. 3.10 The 4 V
characteristics of big data

- Variety: There are many types of data. In addition to social networking and Internet search, some sensors are installed on roads, bridges, cars, and airplanes, each of which increases the diversity of data. The diversity of types also allows data to be divided into structured and unstructured data. Unstructured data, including pictures, audio, video, and geographic location information, is more and more popular compared with structured data which is easy to store in the past. These multi-types of data require more processing power.
- Value: Low-value density. The value density is inversely proportional to the size of the total data. That is, to dig a valuable piece of information, it is often necessary to find from a growing amount of data. Take consumer shopping in e-commerce as an example. In the past, it was possible to analyze and judge consumer purchasing behavior from 100,000 commodity browsing data, but now it is possible to mine valuable information from tens of millions or more commodity browsing data. How to use powerful deep complex analysis (machine learning, artificial intelligence, etc.) to more quickly complete the value "mining" of data and to predict future trends and patterns has become an urgent problem to be solved in the current context of large data.
- Velocity: Fast processing. This is the most significant feature that distinguishes large data from traditional data mining. In front of the huge amount of data, the efficiency of data processing is the life of enterprises. Enterprises need to know not only how to quickly create data, but also how to quickly process, analyze, and return to users to meet their real-time needs.

3.6.2 Increasing Data on Manufacturing

With the deep integration of the new generation of information technology and manufacturing industry, the gestation and development of industrial 4.0 make information technology penetrate into every link of the industrial chain of manufacturing enterprises, such as barcode, two-dimensional code, RFID, and other IoTs identification. Industrial automatic control systems such as industrial sensors, PLC, and software technology such as ERP, CAD/CAM, and MES have been widely used in manufacturing enterprises. Manufacturing enterprises operations are also increasingly dependent on information technology. As a result, both the entire value chain of manufacturing industry and the entire life cycle of manufacturing products involve a lot of data, and the data of manufacturing enterprises also shows an explosive growth trend (Fig. 3.11).

The revolutionary nature of industrial 4.0 is that it no longer starts with the productive forces demand at the manufacturing side; instead it takes the client value as the starting point of the entire industrial chain, which changes the previous model of industrial value chain from the production side to the consumer side, upstream to downstream. It provides customized products and services from the value demand of the client side and uses this as the common goal of the entire industrial chain. It makes all links of the whole industry chain achieve synergistic optimization.

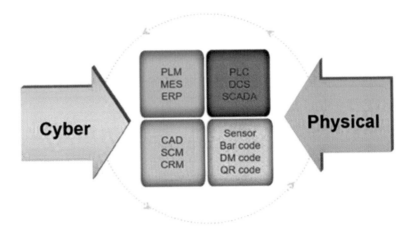

Fig. 3.11 Industrial big data generated by CPS

The basis of synergistic optimization is to visualize production data and make production decisions using large data analysis.

The production line is running at a high speed. The amount of data generated, collected, and processed by the production equipment is much larger than that generated by computers and human in the enterprise, and the real-time requirement of the data is also higher. On the production site, data is collected every few seconds, which can be used for a lot of analysis and mining, including device startup rate, spindle operation rate, spindle load rate, operation rate, failure rate, productivity, comprehensive utilization rate of equipment, parts eligibility rate, quality percentage, and so on.

In the aspect of production process improvement, using these big data, the whole production process can be analyzed, and the execution of each link can be controlled in real time. Once a process is found to have deviated from the standard process, an alarm signal is generated, which enables faster detection of errors or bottlenecks and easier resolution of problems.

In the aspect of optimizing production process, using large industrial data, a virtual model can be established for the production process of products, and the production process can be optimized by simulation. When all process data can be built in the network system, it will help manufacturing enterprises improve their production process by means of visualization.

In the aspect of energy consumption analysis, the sensor is used to centrally monitor the production process of all production equipment, and find abnormal or peak energy consumption situations, so as to optimize the efficiency of energy use and reduce energy consumption.

Big industrial data is the necessary means to accelerate technological innovation, service innovation, and management innovation in manufacturing industry. Through the specific application of large industrial data, data-driven production mode can be

Fig. 3.12 Application value
of industrial big data

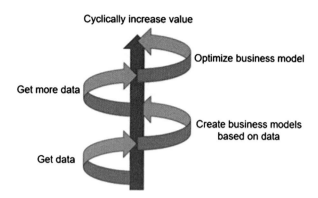

achieved, which makes the processes inside and between the cooperative factories more agile and efficient, thus promoting and implementing the innovation of products, processes, operations, and business models (Fig. 3.12).

3.6.3   Big Data Provide Basis for Mass Customization

Big data is the basis of manufacturing intelligence. Its application in manufacturing mass customization includes data collection, data management, order management, intelligent manufacturing, customization platform, etc. Its core is customization platform. Custom data up to a certain order of magnitude enables large data applications, through the mining of large data, to achieve more applications such as popular prediction, accurate matching, fashion management, social applications, marketing push, and so on. At the same time, big data can help manufacturing enterprises to improve the targeted marketing, reduce the cost of logistics and inventory, and reduce the risk of input of production resources.

Using big data for analysis will result in a significant efficiency increase in warehousing, distribution, and sales, as well as a significant decrease in costs. This will greatly reduce inventory and optimize the supply chain. At the same time, manufacturing enterprises can accurately predict the demand for goods in different market areas around the world using data such as sales data and sensor data. Because inventory and sales prices can be tracked, manufacturing companies can save a lot of costs.

To achieve the personalized needs of consumers, on the one hand, manufacturing enterprises need to be able to produce products or services that meet the personalized preferences of consumers; on the other hand, the Internet is required to provide the personalized customized needs of consumers. Due to the large number of consumers and the different needs of each person, the specific information of the needs is different. In addition, the ever-changing needs constitute the large data of product requirements. The interaction and transaction between consumers and manufacturing enterprises will also produce a large amount of data. Mining and analyzing these consumer dynamic data can help consumers participate in innovative activities such as product demand analysis and product design, and contribute to product innovation. Manufacturing enterprises process these data and then transfer it to smart devices, data mining, equipment adjustment, raw material preparation and other steps to produce customized products that meet personalized needs.

The personalization of consumer demand requires traditional manufacturing industry to break through the existing production mode and manufacturing mode and to process and mine large amounts of data and information according to the needs of consumers. At the same time, in the production process of these non-standardized products, a large amount of production information and data are generated, which needs to be collected, processed, and analyzed in time to guide production in turn.

These two big data information streams are ultimately transmitted between smart devices through the Internet. Smart devices analyze, judge, decide, adjust, control, and continue to carry out smart production, producing high-quality personalized products. In a sense, big data is also the basic element of smart manufacturing.

Big data in the process of smart manufacturing is generated by the interaction and fusion of "cyber" and "physical" worlds. Big data applications will bring a new era of innovation and change for manufacturing enterprises. Based on the traditional information data of manufacturing production management, through physical data perception brought by IoTs, a private cloud of production data in the era of "co-networking" has been formed, which innovates the research, development, production, operation, marketing and management methods of manufacturing enterprises. These innovations bring faster speed, higher efficiency, and higher insight to manufacturing enterprises.

Intelligent data has great potential value. With more and more machines and devices joining the industrial Internet, the synergistic effect of machines and instruments across the whole set and network can be achieved. Intelligent data has the following values:

- Optimize network: A variety of devices or machines that are interconnected within a network system can collaborate with each other through the Internet to improve the overall operational efficiency of the network. For example, in the medical field, medical data are interconnected, transmitted seamlessly to medical institutions and patients. It not only reduces patients' waiting time and helps patients to use the correct medical equipment more quickly but also makes the medical equipment more efficient and the quality of medical services better. The great value of intelligent data can also be reflected in routing optimization in transportation networks. When many vehicles are interconnected, they will know their location and destination, as well as the location and destination of other vehicles in the network system, allowing optimized routing to find the most effective artificial intelligence solution.

- Optimize operation and maintenance: Intelligent data enables optimization and low cost, and facilitates the operation and maintenance of the entire device or machine. For example, after networking machines, components, and links, a monitorable device status will be achieved. The optimal number of parts can be delivered to the exact location at the right time, the inventory requirements and maintenance costs of parts will be reduced, and the stability of the equipment or machine will be improved.
- Recovery system: Help network systems recover more quickly and effectively after a devastating attack by building extensive big data information. For example, in the event of an earthquake or other natural disasters, a network of smart meters, sensors, and other smart devices and systems can be used for rapid detection to isolate the faulty device or machine from a larger scale of faults due to series connection.
- Autonomous learning: The operating experience of each device or machine can be aggregated into a big data, enabling the entire device or machine to learn independently. This way of self-learning is not possible on a single machine. For example, data collected from many aircraft plus historical location and flight data can provide information about the performance of aircraft in a variety of environments. As more and more machines are connected to a system, the result of producing an infinite number of data-only systems will be an ever-expanding network system that can learn independently and become increasingly intelligent.

Once a large amount of intelligent data has been collected by industrial equipment, intelligent systems can be used to make intelligent decision-making with business value. The combination of devices, data, network collaboration, and real-time update will benefit many industries.

GE predicts that airlines will lose more than $40 billion a year from flight delays. Ten percent of the delays were due to a lack of maintenance for the aircraft. Meanwhile, the global aviation industry spends $170 billion a year on fuel, and 18–22% of this fuel consumption is wasted, according to the International Air Transport Association (IATA). GE's industrial Internet reduces delays by 1,000 times a year by monitoring and counting data from the aircraft transport administration and spare parts systems to analyze problems with maintenance. At the same time, choosing the right time for maintenance can also reduce the cost of investment in equipment. Through shipping data, the path to reduce fuel consumption is explored, which optimizes flight scheduling by reducing energy consumption by 2%, saving $20 million annually and reducing large amounts of carbon dioxide emissions.

Healthcare accounts for 10% of global GDP and is a sizable industry. GE predicts that the inefficiency of the medical sector will result in $731 billion of waste annually, especially in clinical care, which accounts for 59% and reaches $429 billion. The information asymmetry between medical practitioners and medical devices is the main reason. For example, processes such as nurse dressing changes, magnetic resonance imaging, and doctor diagnostics cannot be shared in real time. The medical professionals and medical devices will be networked, information such as diagnosis, surgery, and prescription will be shared, and network collaborative diagnosis and treatment will be developed. GE's industrial Internet can reduce the cost of medical devices by 15–30% and improve the efficiency of medical practitioners by managing each hospital bed, workflow of each diagnosis, handling, and communication of medical devices. By improving medical practitioners, business processes, and devices, the time saved can be as much as 15–20% of patients.

3.7   Intelligent Robots: The Main Force of Industrial 4.0

3.7.1   Why Do We Need a Large Number of Industrial Robots for Intelligent Manufacturing?

Industrial robots were first applied on a large scale in the production line of automotive manufacturing and then developed manufacturing countries such as Japan, Germany, and the United States began to use robots in other industrial production. Since the twenty-first century, with the continuous increase of labor costs and technology, the transformation and upgrade of manufacturing industry has been carried out successively in various countries, and the upsurge of robot replacing human has emerged.

In addition to costs reducing, industrial production with robots has a series of advantages, such as significantly increasing production efficiency, improving product yield, ensuring product quality, and enhancing production flexibility (Fig. 3.13).

- Industrial robots can do some monotonous, frequent, and repetitive long-term jobs on behalf of people. It only eliminates dull and tasteless work but also reduces the labor intensity of workers.
- Industrial robots can be widely used in hazardous and harsh environments, such as stamping, pressure casting, heat treatment, welding, coating, plastic product shaping, mechanical processing, and simple assembly.
- Industrial robots can complete the handling or technological operation of hazardous materials, enhance the health and safety of the workplace, engage in work under special circumstances, and reduce labor disputes.
- Industrial robots can improve the degree of automation, reduce the pause time in the process, and thus improve the efficiency of production.
- Industrial robots can improve the processing capacity of parts and components, ensure product quality, and improve product yield, thereby improving product quality. It is an effective scheme for enterprises to supplement and replace labor force.
- Industrial robots can improve automation production efficiency, facilitate adjustment of production capacity, and achieve flexible manufacturing.

A simple and repetitive work

(Single program control)

A variety of complex and diverse work

(Deep learning through cloud computing and artificial intelligence)

Fig. 3.13 Industrial robot toward intelligence

There are three main necessities for industrial robots to be used in manufacturing (Fig. 3.14).

3.7.1.1 Basic Needs

Robots are more adaptable to high-intensity, repetitive, and harsh environment jobs and also the best choice to fill the labor shortage. At present, industrial robots can replace human beings to do most of the work in manufacturing industries such as sorting, handling, welding, mechanical processing, assembly, testing, and stacking. In particular, the rising cost of labor in manufacturing and the falling price of robots have enhanced the cost-effectiveness of robotic industrial applications. Take China as an example, it has already facing the severe situation of insufficient labor force. The surplus rural labor force is gradually decreasing, and the Lewis inflection point that the surplus labor force turns to shortage is coming. At the same time, the aging

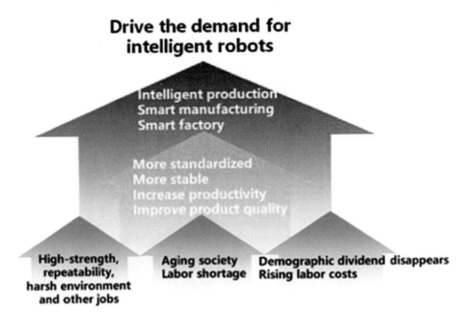

Fig. 3.14 Drivers driving the demand for smart robots

trend of the population structure, with the number of years of declining dependency ratios (the proportion of people who are not working beyond the age of 15–64 years) coming down to a low point and then starting to rise, will be followed by the gradual disappearance of the "population dividend." With the approaching of Lewis inflection point and the disappearance of population dividend, the rapid rise of labor costs in manufacturing industry has greatly squeezed the original slim profit space of small- and medium-sized enterprises, driving the accelerated development of the demand for industrial robots from the basic needs.

3.7.1.2 Requirements for Development

The transformation and upgrading demand of manufacturing industry for product quality and productivity improvement is also driving the accelerated development of industrial robots. With low added value of products and high labor costs, it now compresses the profit space of processing enterprises. Cost forces manufacturers to transform to automatic and efficient production mode. Using industrial robots for production is more standardized, more stable, and more guaranteed for production efficiency and product quality. With the progress of technology, the functions of industrial robots are becoming more and more powerful. The traditional technical indicators such as degree of freedom, precision, operation range, and carrying capacity to measure the level of industrial robots have been significantly improved. Before 2000, 6-axis robots were also the synonym of high-end industrial robots. At present, 6-axis industrial robots are very popular. Many high-end robots have more than six axes. More degrees of freedom have significantly improved the flexibility of the robot, which is no longer limited to simple repetitive work. For example, Epson uses a robotic hand to assemble watch parts, suggesting that robots can do not only simple manufacturing jobs but also and complex and sophisticated jobs.

In addition, with the development of the precision decelerator, which is the core component of the robot, the precision of industrial robots has been greatly improved over 10 years ago, and the operating range, maximum working speed and carrying capacity have also been significantly improved.

Therefore, from the function and actual technical level point of view, industrial robots can completely replace workers to do most repetitive jobs, and free human from heavy physical work.

3.7.1.3 General Trend

Manufacturing industry is the main application field of industrial robots, where they are widely used in production process automation. At present, a large number of robots are used in the large-scale production technology of automotive industry and electronic manufacturing industry, and they will be used in various industries of manufacturing industry in the future.

3.7.2 Scenarios for Industrial Robots

Industrial robot is a production equipment, whose advantages are to improve production efficiency, reduce production costs, and ensure product quality. Therefore, it is widely used in a variety of scenarios. For example, industrial robots are commonly used in welding, painting, assembly, handling, transportation, detection, and inspection.

3.7.2.1 Welding Robot

The welding process mainly produces heat by discharging and uses melted electrodes to connect the plates. The discharge process produces ultraviolet light and toxic gases. Welders usually hold baffles to perform welding work, but there is still be more or less personal injury. Industrial robots have been used early in the field of welding, which effectively freeing welders from toxic environments.

At present, with the development of various functions, welding robots can completely replace skilled welders. The three robots coordinate the implementation of the welding system. The central robot performs the welding. The two-side robots adjust the working angle of the welded objects and complete the welding work conveniently with the central robot.

3.7.2.2   Painting Robot

The paint spray robot is an industrial robot that can automatically spray paint, which is generally integrated in the system manufacturing as a unit of the spray production line.

3.7.2.3   Assembly Robot

In the past, most of the assembly uses were mainly concentrated in the electronic information manufacturing industry, mainly used to assemble electronic components on printed circuit boards. In recent years, with the improvement of the accuracy of industrial robots, some complex assembly jobs that must rely on manual operation have been gradually replaced by industrial robots.

The arm assembly robot pictured above has a left-handed control screw and a right-handed electric screwdriver to tighten the screw. Completely the same as the worker's way of working, it is expected to be widely used in various assembly jobs in the future. Double-arm robots can do things that traditional robots cannot do, such as fine assembly. It is in line with the development of robots to a more flexible direction.

3.7.2.4   Handling Robots

Handling robots are industrial robots capable of automated handling. The carrying robot can install different end-effectors to carry workpieces of different shapes and states, which greatly reduces the worker's heavy physical labor. It is widely used in the automatic handling of machine tools, feeding and unloading, automatic production lines for stampers, automatic assembly lines, stacking and handling, containers, etc.

3.7.2.5   AGV Robot

AGV robot is an industrial robot that automatically carries goods and realizes the function of logistics in the workshop. Portable AGV is widely used in mechanical, electronic, textile, paper, cigarette, food, and other industries. The main features are: AGV is a mobile conveyor, which does not occupy fixed ground space; high flexibility, easy to change the operation path; high system reliability, even if one AGV fails, the whole system can still operate normally. In addition, AGV system can easily connect with management system through TCP/IP protocol, which is recognized as the construction of unmanned workshops, automated warehouses, to achieve self-logistics. For example, in automotive production line, AGV can be used to achieve dynamic automatic assembly of engine, rear bridge, fuel tank, and other components. In large LCD panel production line, automatic assembly can be achieved through AGV, which can greatly improve production efficiency.

3.7.2.6 Detection and Inspection Robot

The consumer goods industry is also an application market for industrial robots. For example, the pharmaceutical industry's drug detection and analysis processing robots can replace testers for drug testing and monitoring analysis. With the use of robots, it is more accurate than skilled testers, and the accuracy of data collection is higher, which can achieve better experimental results. At the same time, it can effectively replace the tester in some dangerous work environments where virus samples are detected.

3.7.3 The Latest Application Cases of Industrial Robots in Intelligent Manufacturing

3.7.3.1 BMW: Robots Take Over Factory

The popularity of "iconic hardware" in the era of industrial revolution, robots has greatly liberated workers. According to the World Robotics 2019 published by the International Federation of Robotics, Germany has an average of 328 robots per 10,000 workers.

German automotive manufacturing equipment is far more advanced and intelligent than we think. There are only a few workers in the whole workshop; instead a lot of robots are busy regularly, and they are next to each production line, which is just like the factory of the future. The production process is divided into many very small segments, each segment is processed according to the computer program settings, strictly following the established sequence, and the segments are linked with high-precision automatic transmission mechanism to achieve flexible production and shorten the production cycle.

In fact, the robot is taking over on a large scale of the BMW Tiexi plant in Shenyang, China. According to the data, there are 642 robots in the body shop of this factory at present, each of which has its own specific work responsibilities. They are busy and professional on different production lines. Looking down the stairs, there are hardly to see any worker in the whole workshop.

It is said that German manufacturers have a deep-rooted idea that workers can never avoid making mistakes. For this reason, they think of breaking every process into small tasks that machines can perform so that never-failing robots can do them. That is, in the future, factories will be completely manufactured by machines themselves, while the role of people is simply to make programming, issue production instructions, and maintain the efficient and reliable operation of production lines.

3.7.3.2   Kuka Robots: Robots Produce Robots

Driven by Industrial 4.0, industrial robots will promote the transformation of production and manufacturing to a flexible and personalized direction. Advanced and flexible fully automated production requires that robots be fully integrated into the production process. This presents an important opportunity for Kuka, Germany's largest robotics company.

Kuka's products are robots, and their own production lines are also using robots in production, so the phenomenon of "robots produced by robots" was born. Kuka staff once said that the industrial robots produced by Kuka are their first customers. With advanced robot manufacturing technology, Kuka has achieved a high degree of automation of production, and the entire factory can see huge arms waving everywhere, but few human beings exist.

Reference

1. J. Akhtar, "Production planning and control (sap pp) with sap erp (2nd edition)," *SAP Press*, 2016.

Chapter 4
5G Communication Technology in Industry 4.0

This chapter introduces techniques used in 5G communication technology. After understanding what Industry 4.0 is, it is obvious that real-time communication between cloud computing platforms and factory production facilities in smart manufacturing processes is very important. The information interaction between massive sensors and artificial intelligence platforms and efficient interaction with human-machine interfaces have diverse requirements and are extremely demanding on performance for communication networks and the need to introduce highly reliable wireless communication technology.

Compared to wired communication systems, wireless systems can obviously break through technical and economic limitations in a short time. In the future, a huge number of low-cost sensors and actuators will be used in industrial production. There is no other way but to use effective wireless systems. Both Bluetooth and WLAN (wireless local area networks) are developing low-cost structures, but so far there are only very limited mechanisms to solve problems such as anti-interference, information security, and response time.

From the perspective of the application of high-reliability 5G mobile communication technology in Industry 4.0, on the one hand, wireless manufacturing equipment makes modular production possible. On the other hand, because wireless networks can make the construction and reconstruction of factories and production lines more convenient, it can reduce a lot of maintenance work to reduce costs through wireless.

4.1 eMBB to Achieve Industrial Internet

The mission of Industry 4.0 is to wireless networking and intelligentizing of the factory. The fourth revolution in the world, represented by Industry 4.0, is intensifying. Manufacturing elements such as machines, equipment, people, and

© Springer Nature Singapore Pte Ltd. 2020
X. Wang, L. Gao, *When 5G Meets Industry 4.0*,
https://doi.org/10.1007/978-981-15-6732-2_4

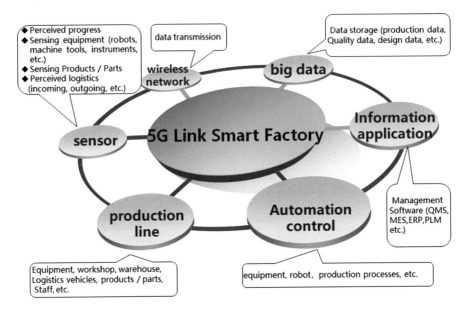

Fig. 4.1 5G links smart factory

products are no longer independent individuals. They are closely linked through 5G to achieve a more coordinated and efficient manufacturing system (Fig. 4.1).

Smart factories are one of the important application scenarios of 5G technology. The 5G network is used to seamlessly connect production equipments and further open up design, procurement, warehousing, logistics, and other links to make production more flat, customized, and intelligent, so as to construct a future-oriented intelligent manufacturing network. In the future, the production of smart factories involves the judgment and decision-making of logistics, material loading, storage, and other schemes. 5G technology can provide cloud computing network platform for smart factories. Precision sensing technology acts on countless sensors and reports information status in a very short period of time. A large amount of industrial big data is collected through 5G networks, and a huge database begins to form. Industrial robots combine super-computing capabilities of cloud computing for autonomous learning and accurate judgments to provide the best solution. In some specific scenarios, using D2D (device-to-device) technology under 5G, objects communicate directly with objects, which further reduces the end-to-end delay of services and achieves shunted in network load. At the same time, the response is more agile.

Data is the lifeblood of smart factories. Based on systematic analysis, the data will help promote the smooth development of processes, detect operational errors, and provide user feedback. When the scale and scope reach a certain level, the data can be used to predict the inefficiency of operations and asset utilization, as well as changes in procurement and demand. Data inside smart factories can exist in

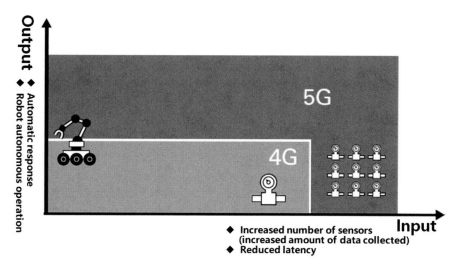

Fig. 4.2 Massive data for smart factories

many forms and be used for a variety of purposes, such as discrete information related to environmental conditions, humidity, temperature, and pollutants. The way of data collection and processing and the corresponding actions based on data are the key to the value of data. To achieve the effective operation of smart factories, manufacturing companies must use 5G's eMBB technology to continuously create and collect data streams, manage and store a large amount of information, analyze the data in a variety of potentially more complex ways, and take corresponding actions based on the data (Fig. 4.2).

In addition, there is human-computer interaction in intelligent manufacturing systems, that is, the interaction between humans and robots. There is also the use of AI to drive and optimize products and processes. Factories need to do some predictive maintenance or predict the energy consumption of machines, etc. More and more of these functions can be implemented in smart factories.

- Ubiquitous connection. It has the ability to comprehensively collect data on various production factors such as equipment, software, and personnel.
- Cloud services. It is to realize massive data storage, management, and computing based on cloud computing architecture.
- Knowledge accumulation. It can provide data analysis capabilities based on industrial knowledge mechanisms and achieve only solidification, accumulation, and reuse.
- Application innovation. It can call platform functions and resources, provide open industrial APP development environment, and realize innovative application of industrial APP.

With eMBB connection speed of 5G networks, smart factories have become an application platform for various intelligent technologies. In addition to the application of the above four types of technologies, smart factories are expected to combine with many advanced technologies in the future to maximize resource utilization, production efficiency, and economic benefits. For example, with the help of 5G high-speed network, it can collect the related energy efficiency data of key equipment manufacturing, production process, energy supply, etc. and use energy management system to manage and analyze the related energy efficiency data. In addition, it can timely discover the fluctuations and abnormalities of energy efficiency and adjust the production process, equipment, energy supply, and personnel accordingly on the premise of ensuring normal production, so as to improve the energy efficiency of the production process. Furthermore, it can use ERP for raw material inventory management, including various raw material and supplier information. When the customer order is placed, ERP automatically calculates the raw materials needed and the purchase time of raw materials according to the supplier information, so as to ensure the lowest or even zero inventory cost while meeting the delivery time.

Therefore, the smart factories in the 5G era will greatly improve labor conditions, reduce manual intervention in the production line, and improve the controllability of the production process. The most important thing is to get through all processes of the enterprise with the help of information technology; achieve the interconnection of all links from design, production, to sales; and achieve the integration and optimization of resources on this basis, so as to further improve the production efficiency and product quality of the enterprise.

In the era of traditional manufacturing, materials, energy, and information are the three elements of factory production. The history of the development of traditional manufacturing is the history of factories using materials, energy, and information for material production. Any technological revolution in the fields of materials, energy, and information will inevitably lead to a revolution in the mode of production and a leap in the development of productivity. However, with the development of mobile Internet, cloud computing, big data technologies, and the evolution of mobile terminals such as computers to smartphones, more and more powerful and intelligent devices have wirelessly achieved interconnection with the Internet or devices. From this, the IoTs, the service Internet, and the data network are derived, driving the integration of the physical world and the information world in the form of a CPS. It can also be said that this technological progress has enabled the interconnection of resources, information, goods, equipment, and people in the manufacturing field.

In the future, through interconnection, cloud computing, big data, and other new Internet technologies, combined with previous automation technologies, production, and processing, will achieve vertical system integration. Cooperation between production equipment, workers, and equipment will link the entire factory to form a CPS, which can not only cooperate and respond to each other but also carry out personalized production and manufacturing. Eventually, it can adjust the productivity of products, adjust the amount and size of resources, and adopt the most resource-saving way.

In addition, the working range of the automation control system and sensor system in the factory can be either hundreds of square kilometers or tens of thousands of square kilometers, and it may even be distributed deployment. According to different production scenarios, there may be tens of thousands of sensors and actuators in the production area of the manufacturing plant, which requires a massive connection capacity of the communication network as a support.

Industry 4.0 has long defined its own top-level architecture, RAMI (Reference Architectural Model Industrie), which describes the framework of Industry 4.0 from three dimensions, representing Germany's global thinking on it. With this model, various enterprises, especially SMEs, can find their place in the entire system. This is an unprecedented confluence. Everything, including human-machine, hardware and software, and virtual and reality, must be connected. A new concept of communication is needed to achieve this integration. From the practice of industrial giants such as Siemens, it is basically based on such a communication framework to build a global domain (Fig. 4.3).

Although most of people misunderstand that the real-time performance of the communication is the key. The bigger challenge of communication, which is a vital subsystem in Industry 4.0 systems, is to seamlessly integrate all the networks covered by the communication requirements. These networks include from the local control network with strict deterministic requirements, the operation and

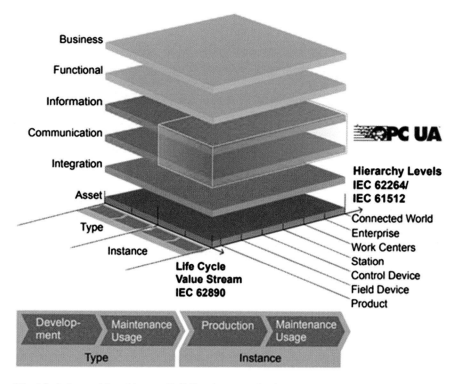

Fig. 4.3 Industry 4.0 architecture RAMI and communication. (The source of this figure is from the Mechanical Engineering Industry Association)

management network inside the factory, the planning and scheduling network of the enterprise production, and the cross-enterprise communication with global connections, self-configuration based on network communication, and integration of various network management.

To pass from the underlying sensors to the cloud, data needs to have a diverse path. From the floor of the factory to the ceiling of the company's executives, the network that traverses all the way and the data connecting the pipelines. How to achieve a seamless integration of communications becomes the biggest challenge.

Achieving a smart factory requires next-generation information technologies such as cloud computing, big data, IoTs, AI, and especially a 5G network. In the 5G era, smart factories will greatly improve labor conditions, reduce manual intervention on production lines, and improve the controllability of the production process. The most important thing is to use information technology to get through various processes of the enterprise and achieve interconnection from design, production, to sales. And, on this basis, the integration and optimization of resources are achieved.

Therefore, with the deep integration of the new generation of information technology and manufacturing industry, many countries and enterprises have realized that the industrial Internet is the top-level ecosystem representing the direction of the deep integration and innovation of the new generation of information technology and manufacturing industry. The future industrial Internet platform includes not only hardware fields, such as production equipment, production materials, and production products, but also various management software, data, and service fields. With the deep integration of information and communication technology and manufacturing industry, the trend of industrial production transforming and upgrading to digitalization has become increasingly obvious. Digitalized knowledge and information data have become the key production factors.

The Predix Industrial Internet Platform launched by General Electric (GE) is an industrial operating system, and there are many modules that can be used by various companies to build their own solutions based on their industry background. Predix is mainly a three-tier architecture. It is the first industrial big data-based cloud platform for the industrial field. The bottom is the IaaS layer that provides infrastructure services, the middle is the PaaS layer, and the top is the software and service layer SaaS. Predix utilizes these three layers of cloud computing architecture to connect various industrial equipment and suppliers with each other to provide APM and operation optimization services. It monitors and analyzes 50 million pieces of big data sent back from tens of millions of sensors on trillions of equipment assets every day, helps customers optimize resource allocation and business processes, reduces risks, and achieves 100% error-free operation.

GE opens this platform to its industrial partners and is expecting to form a huge and perfect ecosystem in the future. Various enterprises will actively develop industrial application software with industrial radiation effect to release, share, and learn from each other and benefit mutually on this platform.

As the orientation of industrial operating system, the platform not only has the ability of management and service for different elements such as equipment, software, users, developers, and data but also considers the basic technology and input-output requirements of the platform. Among them, platform resource management and application services mainly include nine sub-items: equipment access, software deployment, user and developer management, data management, storage computing service, application development service, interplatform call service, security protection service, and new technology service. Furthermore platform basic technology and input-output capacity cover six aspects: platform architecture design, key technology, R&D investment, output benefit, application effect, and quality audit.

The commercialization of 5G has just brought the Internet into the second half, the deepening of the consumer Internet, and the rise of the industrial Internet. 5G emerges at the right time, it develops new applications in the field of consumption and industry, and there will be new forms of business that we still can't imagine, for example, the platformization of manufacturing capabilities, through the modular deployment of data-based manufacturing resources on the platform, can achieve online transactions of manufacturing capabilities (industrial design APP, industrial inspection APP, industrial simulation APP, etc.).

With the deep integration of the new generation of information technology and manufacturing, many countries and enterprises have realized that the industrial Internet platform is the top ecosystem of the future industry and will have a comprehensive, deep, and revolutionary impact on the future industrial development. With the development of the application of the industrial Internet, the network and the entity system will be closely linked, that is, the IoTs will connect the processors and sensors at the production site, so that robots can communicate with each other and form a smart factory.

The future industrial Internet platform will systematically build three functional systems based on network, platform, and security to create a new type of network infrastructure that fully interconnects people, machines, and things, including hardware areas such as production equipment, production materials, production products, various management software, data, and service fields. Through the deep integration of new-generation information technology, such as 5G and manufacturing industry, it will form a new business format and application model for intelligent development. It is beneficial to promote the evolution and upgrade of 5G network infrastructure, promote the leap in the field of 5G mobile communication technology applications from virtual to physical and from life to production, and greatly expand the digital economy space.

The most significant part of the industrial Internet is its cloud computing platform. The massive data generated in industrial production are connected to the industrial cloud platform, and distributed data mining is adopted in a distributed architecture to extract effective production improvement information, which will eventually be used in areas such as predictive maintenance. In the cloud platform, we first get through the data flow and logistics and gather different dimensions within the factory, different stages of the product life cycle, and different actors in the upstream and downstream of the supply chain in the cloud. Secondly, we can use big data and AI technology to analyze and refine the digital analysis model.

Manufacturing intelligence and the industrial Internet have different levels of application scenarios. Firstly, at the enterprise level, it is mainly to improve internal quality and efficiency, reduce costs and inventory, evolve from traditional manufacturing to smart factories, and use data to drive intelligent production capabilities. Secondly, it can achieve the cross-enterprise value chain extension, optimize the cross-enterprise manufacturing resource allocation, and open up the external value chain of enterprises. Finally, it is expected to achieve the ecological construction of the entire industry; drive ecological operation capabilities with data; aggregate the resources of collaborative enterprises, products, users, and other industrial chains; and continue to precipitate, reuse, restructure, and output to achieve the optimal allocation of resources for the entire manufacturing industry.

Manufacturing resources include various types of manufacturing equipment, such as machine tools, machining centers, and computing equipment, and various models, data, software, and domain knowledge in the manufacturing process. Parallel manufacturing should allow everyone involved in product development to instantly exchange information and resources with each other to overcome various problems caused by factors such as the complexity of products and the lack of interchangeable tools due to the different locations of corporate branches, different organizations, and departments. Therefore, in the process of production research and development, for different product objects, different parallel manufacturing methods are used to gradually optimize the manufacturing environment and achieve flexibility and elasticity. In this way, parallel manufacturing requires a large network platform, and "cloud manufacturing" just meets this ecological environment requirement.

Through processing fusion and optimization of data models, platform and transaction deployment and implementation of manufacturing capabilities can share resources to achieve shareable manufacturing resources. Through the big data analysis of the precise docking of supply and demand in manufacturing capabilities, manufacturing resources are dynamically configured to meet demand. The core is the fusion and innovation of human and machine intelligence. The core competitiveness of the industrial Internet platform is reflected in the deep integration of industrial knowledge with big data and AI technology, which accelerates knowledge innovation and value creation.

The function of the industrial Internet platform to cultivate new industrial formats is also gradually emerging in the industrial field. Not only has it spawned new formats for industrial software services, but it has also driven industrial enterprise innovation to form a new service-oriented transformation model.

The first is to promote the industrial big data application ecology. Cloud manufacturing has natural advantages in data collection and data mining. To promote the construction and application of public cloud manufacturing platform based on "unified identification of industrial Internet" is to promote the unified data standards and sharing mechanism of enterprises. Once established, it can greatly accelerate the accumulation of big data of industrial enterprises, accelerate the accumulation of common big data of the industry, and provide important reserves for the further development of industrial big data solutions. Through the analysis and mining of real-time condition monitoring data of products, the maintenance plan of products can be optimized, and new product development can be fed back. On these industrial cloud platforms, product manufacturers, maintenance service providers, product end-users, platform operators, etc. take what they need and cooperate with mutual benefits. It forms an industrial ecosystem with data and services as the core, which also establish a data foundation for the overall development of the manufacturing industry in the future.

The second is to SMEs. In Germany, 90% of companies are SMEs with less than 500 employees. The continuous technological innovation of SMEs, mainly through interenterprise transactions, occupies a large market share in the field of fixed parts and machine tools. The realization of Industry 4.0 requires huge investment in software development, because it is very difficult for SMEs to develop their own. Therefore, German Industry 4.0 is actually based on the industrial structure of its own country. Its purpose is to effectively protect and support domestic SMEs by building a network platform, so that SMEs without the ability to independently develop software can also enjoy technical support. Cloud manufacturing can realize the integration of manufacturing resources and capabilities among enterprises, increase the utilization rate of manufacturing resources and capabilities across the entire society, achieve manufacturing resource and capabilities transactions, support the free trade of manufacturing resources and capabilities within the wide area of SMEs, support SMEs independently publish resource capacity demand and supply information, and achieve free trading of manufacturing resources and capabilities based on enterprise standards. Business collaboration between development, processing, and services among multiple agents, enable SMEs to collaborate with external resources at the level of manufacturing processes.

The third is to realize service-oriented manufacturing and production-oriented service industries. Cloud manufacturing will successfully change the concept of "manufacturing" to "service," transform manufacturing resources into a professional service, borrow the idea of cloud computing, and use information technology to build a public service platform that shares manufacturing resources. The social manufacturing resource pools are connected together to provide various manufacturing services, to achieve open collaboration of manufacturing resources and services, and to highly share social resources. In this way, some manufacturing companies no longer need to invest huge amounts of money to purchase production lines or equipment. They can use the cloud manufacturing platform to "borrow the power and integrate resources" in order to achieve an open value chain and open manufacturing.

The industrial Internet also has a strong permeability, which can be expanded from manufacturing to become an indispensable infrastructure for networked and intelligent upgrades in various industrial fields, to achieve wide connectivity between the upstream, downstream, and cross-domain industries. It also breaks the "information island," promote integration and sharing, fundamentally change the production process, and greatly enhance the relationship with suppliers and customers (Fig. 4.4).

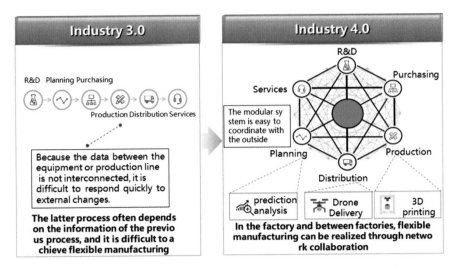

Fig. 4.4 Comparison between Industry 3.0 and Industry 4.0

4.2 mMTC to Realize Digital Twin

In the 2019 Hannover Messe, Germany, known for leading the future of the industry, 5G appeared for the first time in the form of a special exhibition area, and a number of companies conducted centralized display and presentations. One of the main differences between 5G and the previous generations of cellular networks is that 5G can be directly applied to communication between machines and IoTs. Wireless communication including 5G and industrial wifi6 greatly improves the flexibility, versatility, and productivity of smart factories.

Industry is ushering in a beautiful era of communication, and the communication system behind Industry 4.0 is beginning to emerge. With the help of 5G's large-scale link (mMTC) technology, data collected by many sensors, videos, and images captured by cameras can be transmitted in time without compression, which greatly increases the accuracy and speed of identification of information and data.

This is the realization of the so-called digital twin, which fully utilizes data, such as physical models, sensor updates, and operating history, and integrates multidisciplinary, multi-physical quantity, multi-scale, and multi-probability simulation processes to complete mapping in virtual space. Thereby it reflects the entire life cycle process of the corresponding physical manufacturing.

Digital twin is the digital expression of a physical product, so that we can see the possible situation of the actual physical product in this digital product. The related technologies include AR and VR. Digital twins are currently mainly used in industry, where it greatly promotes the changes of product in design, production, maintenance, repair, etc. In the era of Industry 4.0, everything in the physical world will be replicated using digital twin technology.

Smart factories use digital twins, through computer technology, VR, and simulation technology. Digital factories manage and optimize the static and dynamic attributes of digital products, digital production planning, and digital production processes, including establishing the digital environment of factory-level manufacturing systems, to achieve the automation of planning and design in manufacturing system. It is an effective auxiliary means of manufacturing system design and actual production system operation control to build manufacturing system model by using digital factory system and realize the process, planning, and simulation of production and manufacturing comprehensively on computer in digital form.

Smart factories not only focus on the support of the product life cycle but also care about the life cycle of factories and enterprises. Considering from the aspects of planning and design, construction, operation and maintenance, transformation, and upgrading, it comprehensively supports the construction and development of enterprise business based on management, service layer, data layer, physical layer, simulation layers, and other dimensions, so as to achieve economical and reliable optimized operation in the production process.

Traditional smart factories have shortcomings such as models that cannot be shared, lack of dynamic updates, and insufficient intelligent applications. The shortcomings of smart factories can be compensated by the organization of modern smart factories, that is, the organization of digital twins. CPS can achieve the interactive linkage between the information virtual body and the physical entity by constructing a closed-loop channel for the interaction of information space and physical space data. The emergence of digital twins has provided clear ideas, methods, and implementation approaches for CPS. Based on the static model generated by the physical entity modelling, through real-time data collection, data integration, and monitoring, the working status and work progress of the physical entity are dynamically tracked, such as collecting measurement results and tracing information, and the physical entity in the physical space is reconstructed in the information space to form a digital twin with the ability of perception, analysis, decision-making, and execution. Therefore, from this perspective, digital twins are also the core and key technology of CPS (Fig. 4.5).

Smart factories can combine virtual factories with real objects to realize point-to-point information communication between machine workpieces and components. It allows real factories to be expressed in a virtual space and achieves synchronous, consistent, real-time feedback. Digital twin is an integrated application that can be viewed as an application supported by VR. For example, in the past, logistics were two-dimensional. With the support of digital twins, it was completely possible to achieve three-dimensional, where it can use all the space up and down, and improve production efficiency. At the same time, digital twins can improve the statistical efficiency of visual data.

For example, a real-time locating system (RTLS) can guide materials, control mobile robots, monitor workpiece usage, and fully document a real-time positioning system that traces the product assembly. In order to achieve a highly flexible, self-organizing production and logistics solution, it needs to improve the dynamic performance of traditional production and logistics workflow. According

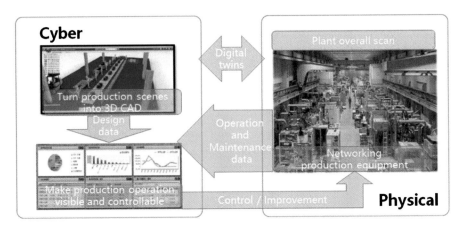

Fig. 4.5 Cyber physical system under the digital twin

to media reports, Siemens centimeter-level SIMATIC RTLS can cover from material warehousing to further processing and final assembly, through the accurate digital mapping of MES systems and business points.

For example, through sensors, RFID, and other IoT identifications, production equipments and products can automatically communicate with each other and collect and summarize production data from smart factories (physical field) into information systems (information field). In the process of assembling and collating these production data, cloud computing mode is mostly used. The manufacturing simulation is carried out by using information technology until the extremely high-precision data information which is completely consistent with the production workshop is obtained. Therefore the design, production planning, logistics, parts procurement, and other links are carried out with the highest efficiency and fastest speed to realize intelligent manufacturing. If this level of smart factory is realized, it can be targeted at high value-added products. Even in small batch production, it can be very flexible and cost-efficient (Fig. 4.6).

4.3   uRLLC to Achieve Concurrent Manufacturing

In smart manufacturing automation control systems, low-latency applications are particularly widespread, such as environmentally sensitive, high-precision production links and chemical dangerous goods production links. In a closed-loop control system of smart manufacturing, the information obtained by sensors, such as pressure and temperature, needs to be transmitted through the network with very low time delay, and the final data needs to be transmitted to the executive components of the system, such as mechanical arm, electronic valve, and heater, to complete the control of high-precision production operation. In the whole process, the network

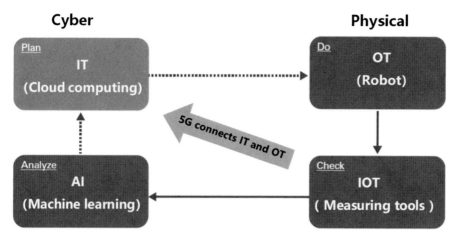

Fig. 4.6 CPS fusion

needs to be highly reliable to ensure the safety and efficiency of the production process.

5G networks can not only provide end-to-end customized network support for smart manufacturing production applications but also achieve end-to-end communication delays as low as 1ms and support 99.999% connection reliability. Strong network capabilities can greatly meet the challenges of delay and reliability in smart manufacturing production, where robots have the ability to self-organize and collaborate to meet flexible production. Smart manufacturing requires devices connected to the cloud through the network, a platform based on ultra-high computing power, and real-time operation and control of the manufacturing process through big data and artificial intelligence. By moving a large number of computing and data storage functions to the cloud, this will greatly reduce the hardware cost and power consumption of the device. In order to meet the needs of flexible manufacturing, robots need to meet the requirements of free movement. Therefore, in the process of intelligent manufacturing, wireless communication networks are required to have extremely low latency and high reliability (Fig. 4.7).

In fact, as early as 2000, ARC, a famous international consulting organization, made a detailed analysis of the development of automation, manufacturing, and information technology for the new development of production and manufacturing mode, together with a comprehensive investigation and research from the perspective of possible impact of the development trend of science and technology on production and manufacturing, and proposed a digital engineering concept described by three dimensions of engineering, production and manufacturing, and supply chain. Through continuous improvement, the collaborative cloud platform of networked manufacturing resources is continuously optimized. Innovative resources, production capacity, and service capacity among enterprises and departments are highly integrated, manufacturing and service operation and maintenance

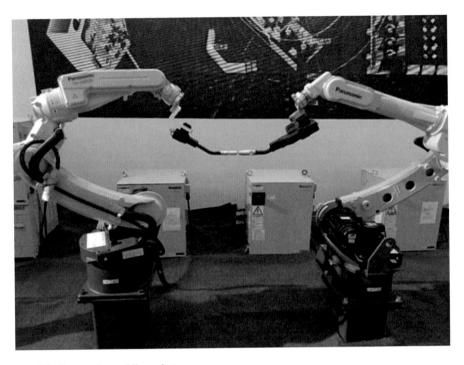

Fig. 4.7 Cooperative welding robot

information are highly shared, and the level of dynamic analysis and flexible configuration of resources and services is significantly enhanced.

Among them, the collaborative manufacturing model (CMM) is formed from the production process management, enterprise business management, to the management of research and development product life cycle. The CMM provides a theoretical basis and effective methods for the transformation of the manufacturing industry. The CMM uses information technology and network technology to organically combine R&D processes, enterprise management processes, and production industry chain processes to form a collaborative manufacturing process, which enables manufacturing management, product design, product service lifecycle, supply chain, and customer relationship management to integrate into a complete closed-loop system of the enterprise. It enables the enterprise's value chain to extend from a single manufacturing link to upstream design and R&D links and the enterprise's management chain to extend from the upstream to the downstream manufacturing control link, so as to form a network collaborative manufacturing system integrating engineering, manufacturing, supply chain, and enterprise management.

In the current era, networked information space and realistic physical space can together form a collaborative space. Information space will have a crucial impact on the development and competitiveness of the manufacturing industry in the future, where it will enter a collaborative era of virtual interaction.

The future form of intelligent manufacturing will be to move manufacturers, component suppliers, sellers, and even consumers online to form a network synergy structure of production resources, human and material resources, and R&D innovation. The main purpose is to achieve the collaboration between market and R&D, the collaboration between R&D and production, and the collaboration between management and communication, so as to form a complete manufacturing network, which is composed of multiple manufacturing companies or participants who exchange goods and information with each other and jointly execute business processes. Enterprise, value chain, and product life cycle, the three dimensions, run through manufacturing participants.

That is, CM (concurrent manufacturing) is realized through network collaboration, where each process in the manufacturing industry will be parallelized, transparent, and flattened to achieve a real intelligent manufacturing. The parallelized intelligent manufacturing process will break through the constraints of the physical world's limited resources by using the unlimited data and information resources in the network world. In this way, you can design and develop, purchase raw materials and components, and organize production and marketing at the same time, thereby reducing operating costs, improving production efficiency, shortening product production cycles, and reducing energy use.

The traditional product production R&D process is generally performed in order. It is a serial development model based on the theory of division of labor by British political economist Adam Smith more than 200 years ago. The theory holds that the finer the division of labor, the higher the work efficiency. Therefore, the serial method is to subdivide the entire product development process into many steps. Each department and individual does only part of the work, and it is relatively independent. After the work is completed, the results are passed to the next department.

However, this serial development mode ignores the communication and coordination between different processes and forms a working environment that focuses on the interests of the vertical departments while does not consider the overall optimization. As a result, conflicts between upstream and downstream cannot be mediated in time, which delays the production and R&D cycle, but also increases the production R&D costs.

To this end, it is necessary to transform the work flow that has precedence over time into simultaneous consideration and simultaneous (or parallel) processing as much as possible. By considering all factors in the whole product life cycle in the product design phase, the production R&D cycle is significantly shortened. Therefore, the designed products not only have good performance but also are easy to manufacture, inspect, and maintain.

In the past, manufacturing enterprises must organize production through five major links, raw materials, equipment, production, transportation, and sales. These five links are relatively fixed and indispensable. In the era of parallel manufacturing, these five links can be relatively independent and become five modules that can be dynamically configured. Each module has its own corresponding software system and its own IoTs sensing system. According to the needs of consumers, the five modules can be efficiently integrated by themselves to meet the manufacturing process requirements. In addition to significantly shorten the construction period, it can also significantly reduce costs.

CPS is becoming the basis for realizing intelligent enterprises and intelligent management in the future industrial system, and it will also be an effective means of integrating various resources and values in the complex and connected world. In addition, it can be said that the complex manufacturing process is also equal to the manufacturing process in the physical space plus that in the information space. The future parallel manufacturing production system will have the ability to make decisions, organize, maintain, and learn independently. It can complete the tasks of precision, flexibility, and intelligence in the production process, as a whole, to achieve a new level of intelligence beyond automation.

The traditional view is that process design work can only be performed after the completion of all product design drawings, production technology preparation and procurement can be performed after all process design drawings are completed, and production technology can be carried out after completion of production technology preparation and procurement. In parallel manufacturing, the related processes are detailed and then crossover in parallel to start the work as soon as possible.

Through the parallel manufacturing on the decentralized value network, product and process design, production technology preparation, procurement, production, and other activities are carried out in parallel. In the process of product development, production, sales, logistics, and service, we should make full use of the means of information and automation. The flexibility is greatly improved. With the help of software and network monitoring the production process can be flexibly adjusted in real time according to the latest situation, instead of completely following the plan of several months or years ago.

4.4 5G LANs

There are many problems with current factories. For example, each production line in the factory uses a different communication protocol. If a production link fails, it cannot be communicated to the next link in time, so that the production efficiency is low. At the same time, the supply chain of some manufacturing enterprises includes many enterprises such as domestic factories, foreign production bases, and customers, which need a large number of external network connections. The realization of mass customization also needs to understand the user's personalized needs from the Internet. The introduction of 5G will effectively solve various problems.

According to media reports, Bosch in Germany has partnered with Nokia to carry out a pilot experiment of factory wireless. Data shows that 5G has lower latency and higher stability than Wi-Fi and 4G. Bosch claims that compared with the large-scale

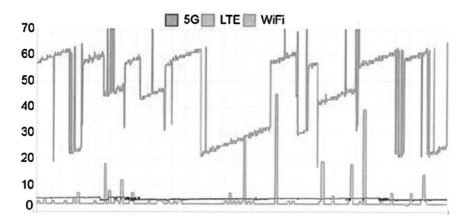

Fig. 4.8 Wireless network connection test of industrial APP. (Source of data: Nokia)

shutdown of factory production lines, the short-term stoppage of some key process lines has greater losses and believes that 5G can provide a low-latency and high-stability communication infrastructure to solve this issue (Fig. 4.8).

Smart factories not only need low latency and high stability but also need to adopt customized network slicing technology based on customer needs. The equipment in a factory usually has different networking requirements based on different uses, purposes, and associated factories. It is very effective to manage multiple different network lines based on the 5G communication environment, which can enable a variety of business needs with differentiated characteristics. In large factories, different production scenarios have different requirements for network service quality. The key to the high-precision process is time delay, and critical tasks need to ensure network reliability and high-speed data analysis and processing. With its end-to-end slicing technology, 5G networks have different quality of service in the same core network and can be flexibly adjusted as needed. For example, the reporting of device status information can be set to the highest service level.

It is difficult to install optical fiber in traditional factories, while Wi-Fi, the past unlimited communication technology, is difficult to expand, and the speed and anti-interference are not strong. In manufacturing facilities, Wi-Fi's frustrating delays can be costly, limiting productivity and reducing profits. In some countries, mobile Internet connections are already more powerful than Wi-Fi. In order to take advantage of higher speeds and reliability, some manufacturers, especially those in the automotive sector, are considering implementing 5G LANs. In addition to 10 times (or more) faster network speed and 50 times less latency than 4G, 5G can support more devices. With the arrival of 5G era, based on the needs of industrial networking, the factory can set up a dedicated 5G network, which also needs a dedicated frequency band.

In the future, factories can cooperate with telecom operators to build their own dedicated 5G LANs using the telecommunications operator's wireless communication technology. 5G LANs will be very useful for businesses that use a large number of connected devices in a centralized area, such as manufacturing facility. For example, the frequency bands allocated by Japan for 5G LANs are two bands: 4.6–4.8 and 28.2–29.1 GHz. Among them, 100 MHz bandwidth of 28.2–28.3 GHz is the current mainly used. Therefore, users can build a 3 Gbps proprietary communication network by themselves. In addition to factories, 5G LANs can also be used in fields such as farm control, security monitoring, and high-definition impact transmission in entertainment venues.

To achieve intelligent manufacturing, before the application of digital, networked, and intelligent software and hardware in manufacturing factories, it is more fundamental to open up the main links of design, production, detection, transportation, storage, distribution, etc. in the production process. The design of the production process contains huge opportunities to improve quality and efficiency, while reduce costs and savings.

The unique concept of 5G LANs is that they enable enterprise customers to run their own local networks through dedicated devices and settings. This approach offers three main advantages:

- Local control: By using dedicated equipment, the dedicated LTE network and its performance are irrelevant to other users, and there is no such problem as sudden traffic surges that may occur in the shared network. This benefit is critical for industrial and enterprise applications, as productivity must be maintained at a high and predictable level. Having local and private networks also gives you complete control over your data. For example, a company can ensure that sensitive data does not leave the house.
- Optimization: By meeting the needs of a single company, a dedicated LTE network can be tailored for that company's specific IoT applications. Examples of such optimizations are quality of service (QoS) and mobility settings. With customized QoS, it can provide consistent services for critical applications regardless of network load. With customized mobile settings, you can optimize behavior for local applications.
- Easy deployment: Use shared and unlicensed spectrum for anyone to use 5G LAN. 5G LAN deployment is very simple, which enables new entities to enjoy LTE and expands the entire LTE ecosystem. In addition, the ability to leverage 5G LANs allows access to functions such as ad hoc networks and network architectures with self-contained, virtual, or hosted core networks.

Compared with other wireless local networks, the main advantage of 5G LANs is that it has a higher capacity to support many devices; at the same time, it has high-bandwidth applications, longer range, seamless mobility, industrial-grade reliability, consistent delay, quality of service, and security. Last but not least is the interoperability among multiple suppliers and 5G roadmap. 5G's roadmap ensures future-oriented solutions with new 5G capabilities, such as NR in 5G, mission-critical services with ultra-reliable and ultra-low-latency communications.

Wi-Fi was originally a wireless network communication within a specified range. The radio wave was limited to a few tens of meters, but it can send and receive large-capacity data without incurring costs, so it is widely used in various places such as enterprises, parks, and stations. However, compared to LTE, there are many security issues. On another side, because communication is sometimes unstable, key procedures are not assured of the full use of Wi-Fi.

In the 4G era, the maximum communication speed is 100 Mbps for download and 50 Mbps for upload, and there is a delay in transmission. Large-capacity data such as video are difficult to send in real time. The 5G LAN not only has the Wi-Fi advantage of "no communication costs" but also can use SIM authentication to achieve security, dedicated high speed, high reliability, and low latency without going through the public network.

5G LANs were initially used mainly for areas with incomplete communications infrastructure and mainly in the fields of mining and energy. However, the demand for driverless cars, robots, and drones is also increasing. For example, German company Bosch is actively adopting 5G LANs, where its electric screwdriver is connected to the network and records the number and sequence of screwing, and experimentally verifying the solution for industrial engineering management. Therefore, the use of electric screwdrivers not only maintains the quality of all manufactured products but also allows real-time reminders of corrections when there is an error in the screwing sequence (Fig. 4.9).

Moreover, the biggest purpose is to feed back all the industrial engineering data to the server. In addition to facilitating industrial engineering management, it can also provide usable information for other products. Bosch previously realized it based on Wi-Fi. Later, it was discovered that 5G LANs are much better than Wi-Fi in terms of anti-band interference, bandwidth (especially uploading), overload, and network security, which is much better than Wi-Fi, and it will be widely used in the future.

From the perspective of improving efficiency and reducing costs, networks in factories also need to be wireless. At the same time, in order to ensure productivity improvements and product quality, the factory is continuously promoting automation.

However, with the expansion of factories and the complication of industrial engineering, 4G is sometimes difficult to handle, and 5G is expected to solve these problems. In the 15th and 16th editions of the 5G standards promoted 3GPP (3rd Generation Partnership Project), eURLLC is specified, and it makes more advanced solutions possible. Deutsche Telekom, one of the founding members of 5G-ACIA, deployed a public network and a 5G LAN in the same network to provide a "dual-slice" network solution. Experiments have been carried out in OSRAM factory of Siemens lighting company (Fig. 4.10).

OSRAM uses 5G LAN for control in industrial robots and AGVs. In some key processes that need to manage a large number of devices, it can also achieve high stability, ultra-low latency, and the same performance as wired networks.

Meanwhile, various data generated by industrial engineering are also collected into edge terminals or cloud computing in real time through 5G LAN, which can be processed through AI to achieve automation (Fig. 4.11).

Audi, the German carmaker, has been using Wi-Fi to control industrial robots in its factories. However, in order to collect data at a higher speed and more flexibility in real time, trials of 5G LAN were begun in 2018. At present, robots used for

Fig. 4.9 Data collected by Bosch based on a dedicated LTE electric screwdriver. (The source of this figure is from Deutsche Telekom)

welding parts have achieved remarkable results. In the next few years, all Wi-Fi in other German factories are expected to be replaced with 5G LANs.

In Germany, 5G has become an important infrastructure for Industry 4.0. In addition to Audi, Daimler, Volkswagen, and other automobile manufacturers, utilities, natural gas, oil, chemical plants, harbor groups, and other enterprises are also considering the construction of 5G network.

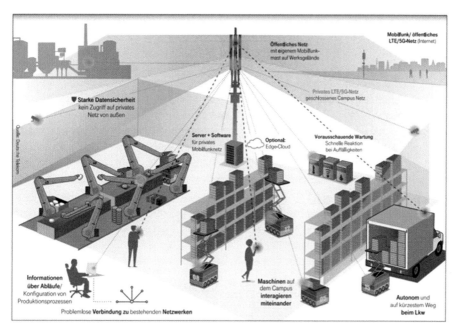

Fig. 4.10 5G experiment at the OSRAM factory. (The source of this figure is from Deutsche Telekom)

Fig. 4.11 AGV with dedicated LTE in Audi factory. (The source of this figure is from Deutsche Telekom)

The lower latency, higher bandwidth, faster speed, and larger capacity brought by 5G wireless technology are driving the digital transformation of manufacturing in various fields, such as AI, AR, predictive maintenance, and collaborative robots. Shortening production cycles makes it more difficult for manufacturers to meet the demand for increasingly complex products while maintaining and complying with quality regulations and standards.

The current transformation of the manufacturing industry can be seen as the integration and promotion of automation upgrades and information technology. This not only is automation and machine substitution but also helps factory to achieve autonomous decision-making, flexibly produce diversified products, and quickly respond to more market changes.

Chapter 5
5G in Real Industrial Scenarios

This chapter describes 5G + Industrial 4.0 scenarios from the perspective of specific applications, for example, the comprehensive perception of production process, the early warning of equipment status, and the guarantee and improvement of product quality.

The ideal smart factory is a flexible system, which can optimize the performance of the whole network, adapt to and learn new environmental conditions in real time or near real time, and automatically run the whole production process by using interconnected IT/OT technology. It can also realize highly reliable operation, minimize human intervention, shorten the time of production and manufacturing, make solutions faster and better, and thereby improve manufacturing efficiency greatly (Table 5.1).

In terms of production, 5G can be combined with related technologies such as ultra-high-definition video, sensors, and control systems to help industrial enterprises achieve remote operation of equipment and real-time monitoring of production processes, which will further improve production efficiency and working conditions for employees. Wireless networks get rid of cable restraints, make mobile applications of robots possible, and enable factories to quickly and cost-effectively switch between different types of product production lines. At the same time, intelligent robots are connected to a cloud-based control center through 5G, and big data and artificial intelligence are used to control the manufacturing process in real time. There are self-organizing and collaborative robots to meet flexible production.

5G meets the requirements for wide coverage, mass connections, and low-cost network connections from warehouse management to logistics distribution and provides tracking and positioning of the product's entire life cycle. In terms of equipment operation and maintenance, 5G can be combined with ultra-high definition to make equipment inspection status transmitted to the cloud for real-time identification in order to improve its quality and efficiency. Experts can provide real-time remote guidance through 5G to reduce equipment maintenance time and costs.

© Springer Nature Singapore Pte Ltd. 2020
X. Wang, L. Gao, *When 5G Meets Industry 4.0*,
https://doi.org/10.1007/978-981-15-6732-2_5

Table 5.1 5G scenarios in industrial applications

Scenes	Produce		Device		Detection	Operating
	Trajectory control	In-plant logistics	M2M control	Remote control	Industrial inspection and measurement	Industrial AR and surveillance
Use case	Components for moving and rotating machine tools or industrial robots	AGV and remote monitoring robot	Multiple controllers/separate machines to assist in completing a function	Under long-distance, harsh working environment, industrial security, and other applications, the application innovation based on cloud manufacturing technology has significantly increased the number of connected terminals	At the same time of production, using industrial cameras to shoot and using AI image recognition algorithm to identify quality problems	Industrial wearable device applications
Demand for 5G	uRLLC	uRLLC	uRLLC	mMTC	uRLLC	eMBB

Video surveillance and machine vision are becoming more and more popular in modern manufacturing enterprises, such as quality inspection, operation and maintenance, and monitoring based on security management behaviors. 5G-based machine vision analysis can now meet a large number of data transmission requirements.

5G can fulfill the network connection and data transmission of massive low-power embedded sensors, as well as unprecedented interaction and coordination between machines, equipment, and people. It makes possible for industrial automation closed-loop control applications (time-sensitive communication) to connect through 5G networks. In the intelligent manufacturing process, AR and other technologies can be used to implement human-machine collaboration, monitor production processes, and step-by-step instructions for production tasks, such as manual assembly process guidance and remote expert business support. In addition, the combination of 5G and industrial AR can be used to train employees to improve their skills.

5.1 Perceivable Production Process

3GPP defines three major scenarios for 5G, each of which is related to manufacturing. eMBB can be used for high-speed mobile broadband services such as AR, VR, and 3D ultra-high-definition video; mMTC can be used for large-scale IoTs such as smart meter reading, smart agriculture, and process automation; uRLLC can be used for unmanned driving, industrial automation control, mobile robotics, remote manufacturing, and other services that require low-latency, high-reliability connections. In all industrial applications, trajectory control should be the most challenging and demanding, which is mainly the application of uRLLC technology.

Automation control is the most basic application in a manufacturing factory, and its core is a closed-loop control system. Each sensor performs continuous measurement during the control cycle of the system, and the measurement data is transmitted to the controller to set the actuator. The typical time for a closed-loop control process cycle is as low as ms level, so the system communication delay needs to reach ms level or even lower to ensure accurate control of the control system, and at the same time, it has extremely high requirements for reliability. If the delay in the production process is too long, or the control information is incorrect during data transmission, it may cause production downtime and huge financial losses.

In addition, in a large-scale production factory, a large number of production links use automatic control processes, so there will be a large number of high-density controllers, sensors, and actuators that need to be connected through a wireless network.

To a certain extent, automation has always been part of the factory, and even high-level automation is not new. However, the term "automation" often refers to the execution of a single and independent task or process. In the past, machine "decision-making" was often based on the linear behavior of automation, such as opening a valve or closing a pump based on a predetermined set of rules. Through the application of AI and the increasing maturity of the CPS, automation increasingly covers the complex optimization decisions usually made by humans.

The main characteristic of the assembly line operation in the Industry 3.0 era is that the operator does not move on the station, while the material is transferred through the assembly line. Instead they simply repeat a fixed action continuously. The advantage is that the operator can avoid labor links such as walking around the workshop and changing tools, thereby significantly improving work efficiency. With the emergence and popularity of PLC (programmable logic controller), automation technology has achieved a major breakthrough. PLC enables some logically complex operations to be performed automatically by the device. At the same time, the development of CNC machine tool technology enables parts to complete a number of complex machining tasks on the machine according to drawings. In addition, the use of industrial robot technology, such as manipulators, also makes it possible to replace simple and repeated fixed operations for operators. Therefore, on the assembly line, operations that are decomposed and composed of standardized actions can be easily completed by automated machines. That is, pipelines can be easily automated. The past 30 years have been the fastest period of globalization. Developed countries have transferred a large number of labor-intensive industries to developing countries with lower labor costs through industrial transfer. For a large number of labor-intensive industries, the overall cost of an assembly line with a high level of automation is often higher than that of a production line with a low level of automation.

However, automated pipelines also have their drawbacks. It cannot be flexibly produced or meet personalized customization, and it is more suitable for manual operation with low repeatability, relatively complicated, and strong perception ability. It is the long-term goal of the manufacturing industry to better meet individual needs and improve the flexibility of production lines.

In Industry 4.0, sensors equipped in production lines and equipment can capture data in real time and then connect to the Internet via wireless communication to transmit data and monitor the production itself in real time. The data of the device sensing and control layer is fused with the enterprise information system to form the CPS, which enables the production data to be transmitted to the cloud computing data center for storage, analysis, and decision-making and in turn to guide the equipment operation. The intelligence of the equipment directly determines the intelligent production level required by Industry 4.0.

In the intelligent manufacturing production scenario, robots are required to have the ability to self-organize and collaborate to meet flexible production, which brings the robot's demand for cloudification. Compared with traditional robots, cloud-based robots need to be connected to a cloud-based control center via a network, based on a platform with ultra-high computing power, and perform real-time operational control of the manufacturing process through big data and artificial intelligence.

Through cloud-based robots, a large number of computing functions and data storage functions are moved to the cloud, which will greatly reduce the hardware cost and power consumption of the robot itself. And in order to meet the needs of flexible manufacturing, the robot needs to meet the requirements of free movement. Therefore, in the process of robot cloudification, wireless communication networks are required to have extremely low latency and high reliability.

5G network is an ideal communication network for cloudized robots, and it is the key to enable clouded robots. 5G slicing networks can provide end-to-end customized network support for cloud-based robot applications. 5G networks can achieve end-to-end communication delays as low as 1 ms and support 99.999% connection reliability. The strong network capability can greatly meet the challenges of delay and reliability of cloud-based robots.

According to media reports, Huawei has cooperated with manufacturing companies in the field of intelligent manufacturing in German. For example, it cooperated with Festo on the 5G slicing network-based cloud robot project. This project tested the high reliability and real-time performance of closed-loop control system of cloud-based robot through 5G uRLLC slicing network.

Production efficiency is the first consideration of manufacturing enterprises. In terms of specific production process, the significance of Industry 4.0 to enterprises is that it can form a closed-loop network of various production resources, including production equipment, factory workers, business management system, and production facilities, and then realize the horizontal and vertical links and end-to-end digitalization of the value chain network throughout the entire intelligent products and systems via IoTs and system service applications integration, so as to improve production efficiency, and finally realize the intelligent factory. Through the horizontal connection of the intelligent factory manufacturing system on the decentralized value network, we can flexibly and timely adjust the production process in the process of product development, production, sales, logistics, and service, with the help of the monitoring and communication of software and network. Instead of completely following the plan of several months or years ago, it can adjust based on the latest situation.

Industry 4.0 connects different devices through data interaction and the CPS, so that the inside and outside of the factory form a whole. And this "integration" is actually to achieve "decentralization" of manufacturing. In Industry 4.0, the production mode is changing from centralized control to decentralized enhanced control, and decentralized production is becoming more flexible than the automated method of the assembly line.

Smart factories are more than simple automation. A smart factory is a flexible system that can optimize the performance of the entire network, adapt to and learn new environmental conditions in real time or near real time, and automatically run the entire production process. Smart factories can operate automatically in the factory floor and continue to develop toward the seamless Internet, such as objects, data, and services (Fig. 5.1).

It is expected that within the next 10 years, 5G networks will cover all corners of the factory. Industrial robots controlled by 5G technology have walked from the glass cabinet to the outside and freely shuttled in the workshop day and night to carry out equipment inspection and repair, feeding, quality inspection, and difficult

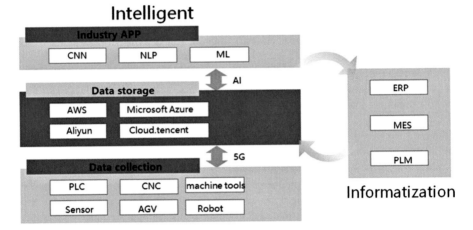

Fig. 5.1 Smart factory-based uRLLC

production operations. The robot becomes a middle-level and grassroots managerial personnel and carries out production coordination and production decision-making through information calculation and accurate judgment. Robots have become senior assistants for humans, replacing humans with tasks that are difficult for humans to complete, and humans and robots can coexist in factories, for example, in:

- Internet of Things: With the promotion of the intelligent transformation of factories, IoTs, as the key supporting technology for connecting people, machines, and equipment, is receiving high attention from enterprises. This demand has also greatly stimulated the development of 5G technology while promoting the application of IoTs.
- Industrial automation control: This is a basic application in a manufacturing factory, where the core is a closed-loop control system. 5G can provide extremely low-time extension, high reliability, and massively connected networks, making it possible for closed-loop control applications to connect through wireless networks.
- Logistics tracking: From warehouse management to logistics distribution, it needs wide coverage, low power consumption, large connection, and-low cost connection technology. In addition, the end-to-end integration of the virtual factory spans the entire product life cycle. In order to connect the widely distributed sold goods, it also needs a low-power, low-cost, and wide coverage network. The horizontal integration within or between enterprises also needs an omnipresent network. 5G network can meet such needs well.

Smart factories represent a leap from traditional automation to fully connected and flexible systems. This system is able to continuously obtain data from interconnected operation and production systems to understand and adapt to new requirements. A true smart factory can integrate physical assets, operating assets, human capital, manufacturing, maintenance, and inventory tracking throughout the system. It is also digitalizing operations through digital twins and other types of activities throughout the manufacturing network. The results may be more efficient and agile for systems, less production downtime, and a stronger ability to predict and adjust changes in the factory or the entire network, thereby further increasing market competitiveness.

Many manufacturing companies have begun to adopt smart factory processes in multiple areas, such as advanced planning and scheduling using real-time production and inventory data, or equipment maintenance using virtual reality technology. However, a true smart factory is a more holistic practice that not only transforms the factory workshop but also affects the whole enterprise and the ecosystem in a larger scope. Smart factories are an integral part of the entire digital supply network and can bring multiple benefits to manufacturing companies, making them more effective in adapting to changing market conditions. Adopting and implementing a smart factory solution can seem complex or even difficult to implement. However, in the context of the rapid development of the technology field and the rapid evolution of future trends, it is almost imperative for manufacturing companies to maintain a market competitiveness or subvert the market competition pattern and shift to a more flexible and adaptive production system. Manufacturing companies must start from a large perspective to fully consider various possibilities, and start from a small place to make controllable adjustments to process methods, so as to quickly promote and expand operations to gradually achieve the vision of building a smart factory and achieve efficiency improvements.

5G network will be the most important feature of smart factory, as well as the most valuable assets. The smart factory must ensure the interconnection of basic processes and materials, so as to generate various data needed for real-time decision-making. In a real sense of smart factory, sensors are all over the assets, so the system can constantly capture data sets from both emerging and traditional channels, ensure the continuous update of data, and reflect the current situation. By integrating data from operating systems, business systems, suppliers, and customers, we can fully control the upstream and downstream processes of the supply chain, thus improving the overall efficiency of the supply network (Fig. 5.3).

The optimized smart factory can achieve a highly reliable operation and minimize manual intervention. Smart factories have automated workflows that allow them to understand asset status simultaneously while optimizing tracking systems and schedules. Energy consumption is also more reasonable, which can effectively increase production, operating time, and quality, reduce costs, and avoid waste.

The data obtained by the smart factory is open and transparent. Through real-time data visualization, the data obtained from processes and products (finished or semi-finished) are processed and transformed into practical insights to assist manual and automated decision-making processes. The transparent network further expands the understanding of equipment conditions and ensures more accurate enterprise decisions through role-based views, real-time warnings and notifications, and real-time tracking and monitoring.

In a smart factory, employees and systems can anticipate and respond to upcoming problems or challenges, before waiting for problems to occur. This feature includes identifying anomalies, stocking and replenishing inventory, identifying and resolving quality issues in advance, and monitoring safety and maintenance issues. Smart factories can predict future outcomes based on historical and real-time data, thereby improving uptime, yield, and quality while preventing safety issues. In smart factories, manufacturing companies can realize digital operations by creating processes, such as digital twins, and further develop predictive capabilities based on automation and integration.

Smart factories have the agility and flexibility to quickly adapt schedules and product changes with minimal impact. Advanced smart factories can also automatically configure equipment and material processes based on products being produced and schedule changes and then control the impact of these changes in real time. In addition, flexibility also enables smart factories to minimize adjustments when schedules and products change, thereby increasing uptime and output, as well as ensuring flexible schedules.

With the above characteristics, manufacturing companies can have a more comprehensive and clear understanding of their assets and systems, effectively respond to the challenges faced by traditional factories, ultimately increase productivity, and respond more flexibly to changing suppliers and customers.

5.2   Early Warning Equipment Status

In the era of Industry 3.0 marked by automation, while reducing labor costs, enterprises have increased the proportion of equipment assets, which makes remote equipment diagnosis and operation face huge challenges. Because no matter how automated the equipment is, it cannot avoid the aging and wear of the equipment. At the same time, the manufacturing industry is an industrial field where the assembly of parts is the main process. Due to the high level of technical complexity of its machining equipment and the difficulty of equipment maintenance, frequent equipment failures and severe damage have restricted the improvement of enterprise equipment management.

Benjamin Franklin, the first scientist and inventor with an international reputation in the history of the United States, once said that 1 percent remedy is far less than 1 percent prevention. It can be seen that preventive maintenance should be achieved in the first place. Through preventive analysis and early warning, on the one hand, it can help maintenance technicians arrange some important preventive maintenance measures in advance to prevent the occurrence of downtime; on the other hand, through intelligent scheduling of preventive maintenance, enterprises can have sufficient time to prepare for equipment upgrades or updates.

Poor maintenance strategies may prevent the factory's overall production capacity from increasing by 5% to 20%. The research also shows that unexpected downtime causes about $50 billion in losses to industrial manufacturers every year and asset failures cause 42% of unexpected downtime. This is a problem for manufacturers, where they have to take the machine offline to ensure that the technology is running at full capacity. Otherwise they will cause the machine to run longer than it should and cause it to fail, which may cause the production stalled for a longer time. Traditional preventive maintenance requires manual numerical calculations and time-out that manufacturers cannot afford. Considering the long time it takes, this method is cumbersome and ineffective.

PricewaterhouseCoopers has said that nearly 100% of manufacturers want to improve efficiency through digital technologies such as predictive maintenance, and manufacturers use machine learning and analytics to improve predictive maintenance, which is expected to increase by 38% in the next 5 years. Combined with 5G wireless capabilities, manufacturers can utilize simpler and more efficient processes for predictive and preventive maintenance.

According to the forecast of GE company, flight delays cost airlines more than $40 billion annually. 10% of the delays were caused by lack of maintenance of the aircraft. At the same time, the global aviation industry's annual fuel costs are as high as 170 billion US dollars (operating income is about 560 billion US dollars), and according to the International Air Transport Association (IATA) survey, 18–22% of these fuel consumption are waste. GE's Industrial Internet analyzes maintenance issues through monitoring and statistics of aircraft and air transportation bureau, which can reduce 1000 delays each year. At the same time, choosing the right time for maintenance can also reduce equipment investment costs. Through the shipping data, it can mine the realization path of reducing fuel energy consumption, so as to optimize the flight scheduling. It reduces the energy consumption by 2%, which saves $20 million cost every year, and reduces a lot of carbon dioxide emissions.

In the traditional industrial production, the inspection of product quality is usually put at the end, which is a kind of post-supervision. Now, after the sensor is implanted into the CNC machine tool, it can collect all kinds of data during the operation of the machine tool in real time, and upload it to the intelligent platform or cloud immediately, and then compare it with the data model continuously. Once it finds any abnormality, it will alarm immediately in a short time.

Recently, Chevron, a chemical giant, also announced its new plan, saying it will use sensors to connect thousands of sets of equipment by 2024, so as to obtain real-time data on the underlying layers of oil field and refinery equipment, know when its equipment needs to be maintained and replaced, and achieve predictive maintenance.

For a company like Chevron that requires a lot of machinery and equipment, the annual maintenance cost is a huge expense. Traditional manufacturers do not know when the equipment will fail, and often use non-deterministic regular maintenance to reduce the risk of machine failure, which causes unnecessary waste. With 5G, enterprises can monitor equipment operation in real time, and troubles can be solved in real time.

In the future smart factory, each production link is clearly visible and highly transparent, and the entire workshop runs orderly and efficiently. In Industry 4.0, the high-bandwidth and low-latency characteristics of 5G networks are used to collect real-time operation data of automation equipment, which enables automation equipment to add certain new functions to the original control functions, so that the intelligent requirements, such as product life cycle management, safety, traceability, and energy saving, can be realized. These new functions added to the production equipment mean that by configuring many sensors for the production line, the equipment has the ability of perception and transmitting the perceived information to the cloud computing data center through the wireless network, and further making the automation equipment have the intelligent function of self-regulation management through big data analysis and decision-making, so as to realize the intelligent equipment and solve all the above problems.

Generally speaking, the remote diagnosis, operation, and maintenance service mainly includes four aspects: establishing standardized information acquisition and control system, automatic diagnosis system, fault prediction model based on expert system, and fault index knowledge base; realizing remote unmanned control of equipment (products), early warning of working environment, operation status monitoring, fault diagnosis, and self-repair; establishing product life cycle analysis platform, life cycle analysis platform of core accessories, and user usage habits information model; and providing intelligent equipment (products) with health monitoring, virtual equipment maintenance scheme formulation and implementation, optimal use scheme push, innovative application opening, and other services (Fig. 5.2).

Specifically, 5G scenarios of remote diagnosis, operation, and maintenance services need to use AI, big data, and automation technologies to achieve the following functions:

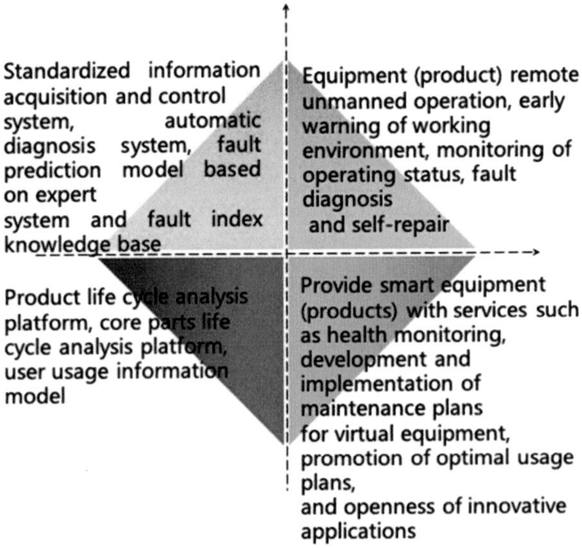

Fig. 5.2 Remote diagnosis, operation, and maintenance services

Standardized information acquisition and control system, automatic diagnosis system, fault prediction model based on expert system and fault index knowledge base

Equipment (product) remote unmanned operation, early warning of working environment, monitoring of operating status, fault diagnosis and self-repair

Product life cycle analysis platform, core parts life cycle analysis platform, user usage information model

Provide smart equipment (products) with services such as health monitoring, development and implementation of maintenance plans for virtual equipment, promotion of optimal usage plans, and openness of innovative applications

- It should be equipped with open data interfaces, which include data collection, communication, and remote control functions. Use the industrial Internet that supports IPv4, IPv6, and other technologies to collect and upload equipment status, operation, environment conditions, and other data, and flexibly adjust equipment operation parameters according to remote instructions.
- It can not only effectively screen, sort, store, and manage the uploaded data of equipment/product but also provide users with daily operation and maintenance, online detection, predictive maintenance, fault warning, diagnosis and repair, operation optimization, remote upgrade, and other services through data mining and analysis.
- It shall achieve information sharing with the product life cycle management system, customer relationship management system, and product development management system.
- It shall establish corresponding expert libraries and expert consulting systems, which can provide intelligent decision support for remote diagnosis of intelligent equipment/products, and propose operation and maintenance solutions to users.
- It should establish an efficient and secure intelligent service system through continuous improvement. The services provided can form real-time and effective interaction with products, greatly improving the integrated application level of embedded system, mobile Internet, big data analysis, and intelligent decision support system (Fig. 5.3).

With the help of AI, remote diagnosis, operation, and maintenance services have been changed from post-maintenance to predictive maintenance. Predictive maintenance mainly depends on big data and AI algorithm. There are two main ideas of AI algorithm: one is based on mechanism discrimination, which is to establish parameter estimation; order determination; time-domain analysis; frequency-domain analysis or multivariable system for unknown objects, linear and

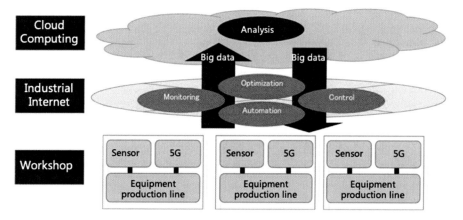

Fig. 5.3 mMTC enhances the remote control capability of equipment operation and maintenance

nonlinear; random or stable system analysis; etc., so as to reveal the internal law and operation mechanism of the system. The other is based on gray-scale modelling, which uses experts system, decision tree, clustering algorithm based on principal component analysis, and advanced machine learning methods to analyze and predict data. Thus, predictive maintenance can effectively reduce equipment downtime, improve equipment utilization, and avoid downtime loss.

In the past, factories were full of machines, and they were next to each other, but the information generated was isolated from each other, rarely "talked" and used. However, in discrete factories or process automation factories, information exchange has a variety of different traffic types and transmission requirements. There is a need for vertical communication between automation control equipment, plant-level systems, and applications, while there is a need for horizontal communication from controller to controller, from controller to field devices (such as actuators, sensors, and drives). Various types of fieldbus and industrial Ethernet supported these different communication needs, thereby dividing the traffic onto separate networks. The formation of such a "data-independent" data governance situation has its historical inevitability, but it is currently becoming a "chain" that restricts the interconnection of all things. 5G will break this limitation, provide solutions applicable to various manufacturing scenarios through network slicing, achieve real-time efficiency and low energy consumption, and simplify deployment, laying a solid foundation for the future development of smart factories.

Firstly, the network slicing technology is used to ensure that the network resources are allocated on demand to meet the requirements of the network in different manufacturing scenarios. Different applications have different requirements for delay, mobility, network coverage, connection density, and connection cost, and more stringent requirements are put forward for the flexible configuration of 5G network, especially for the rapid allocation and redistribution of network resources.

As the most important feature of 5G network, slicing capability combined with a variety of new technologies can help resource be flexibly and dynamically allocated and released to different needs in the whole network. According to the blueprint and input parameters provided by service management, network slicing is created to provide specific network features.

In the process of creating network slices, resources in the infrastructure need to be scheduled, including receiving, transmission, and cloud computing resources. Each infrastructure resource also has its own management function. Through network slice management, according to the different needs of customers, it provides customers with shared or isolated infrastructure resources. Due to the mutual independence of various resources, network slice management simultaneously performs collaborative management among different resources. In the smart factory prototype, the multi-layer and modular management mode is adopted to make the management and cooperation of the whole network slice more general, flexible, and easy to expand.

In addition to critical transaction slicing, 5G smart factories also create mobile broadband slicing and large connection slicing. Different slices share the same infrastructure under the scheduling of the network slice management system, and they do not interfere with each other. Instead, it maintains the independence of their respective services.

Secondly, 5G can optimize network connections and take local traffic offload to meet low latency requirements. The optimization of each slice for business needs is not only reflected in different network functions and features but also in flexible deployment schemes. The deployment of network function modules inside the slice is very flexible and can be deployed in multiple distributed data centers respectively according to business needs. In order to ensure the real-time transaction processing, the key transaction slices in the prototype have high requirements on latency. Functional modules in the user data plane are deployed in the local data center near the end user to reduce the delay as much as possible and ensure the real-time control and response to the production.

In addition, it adopts distributed cloud computing technology to deploy industrial applications and key network functions based on NFV (network function virtualization) technology in local or centralized data centers in a flexible manner. The high-bandwidth and low-latency characteristics of 5G networks have greatly improved intelligent processing capabilities by migrating to the cloud and paving the way for improved intelligence.

Under the connection of the 5G network, the smart factory has become an application platform for various intelligent technologies. In addition to the application of the above four types of technologies, smart factories are expected to combine with many advanced technologies in the future to maximize resource utilization, production efficiency, and economic benefits. For example, with the help of 5G high-speed network, it can easily collect the energy efficiency data of key equipment manufacturing, production process, energy supply, and other links. We can also use ERP for raw material inventory management, including various raw material and supplier information. When a customer order is placed, ERP automatically calculates the raw materials needed, and the purchase time of raw materials according to the supplier information, so as to ensure the lowest or even zero inventory cost while meeting the delivery time.

Therefore, the smart factory will greatly improve labor conditions in the 5G era, reduce manual intervention in the production line, and improve the controllability of the production process. The most important thing is to get through all processes of the enterprise with the help of information technology, realize the interconnection of all links from design, production to sales, and realize the integration and optimization of resources on this basis, so as to further improve the production efficiency and product quality of enterprises.

In terms of system architecture, industrial infrastructure can realize dynamic resource integration and elastic expansion by using network and virtualization technology. Various industrial software can be centrally deployed and maintained, and the control and maintenance of industrial infrastructure can be realized through embedded control system. In terms of business model, the "public cloud" service, which is open to the public, can support users' self-service through application program interface to help complete the business deployment and "private cloud" service, which is used within the enterprise group. It can make fully use of the shared drawings, operating experience, operating data, etc. and form the network collaborative and interconnected manufacturing through the application program interface.

With the support of IoTs technologies such as RFID and sensors, physical equipment (production lines, production equipment, parts, etc.) in the factory will realize the interconnection between things. At the same time, the development of informatization (industrial software, management software, data mining, etc.) also provides the possibility to solve complex manufacturing problems and carry out large-scale collaborative manufacturing. The concept of "cloud manufacturing" came into being.

Cloud applications are often downplayed in terms of function and features due to low throughput, high latency, and inconsistent connectivity of mobile devices. 5G networks can improve the responsiveness, scalability, and flexibility of cloud-based applications. Mobile applications can be more complicated than ever. 5G's ultra-low latency and high throughput will make the cloud computing experience comparable to the desktop LAN connections. It can even enhance today's cloud computing capabilities, which still experience latency and network accessibility issues.

5.3   Improved Product Quality

As the global population is approaching 8 billion and the middle-class consumer group is expanding, it is expected to form a huge market and will have an impact on the consumption layout. The system with customer needs and product "information" functions has become the new core of hardware product sales, and personalized customization has become a trend. In order to meet the diversified and personalized needs of products in different markets around the world, production companies need to update their existing production models, and therefore flexible technology-based production models have become a trend. The flexibility of a system is usually limited by the product family considered during system design. The advent of flexible production has created a need for new technologies.

5G leverages its unique and incomparable advantages to facilitate the large-scale adoption of flexible production. On the other hand, 5G can build a comprehensive information ecosystem centered on people and machines inside and outside the factory and ultimately enable any person and things to share information with each other at any time and any place. When consumers demand personalized goods and services, the relationship between enterprises and consumers changes. Consumers are participating in the production process of enterprises, where they can participate in product design and query product status information in real time through 5G networks across regions.

In actual production practice, 5G scenarios do not always provide centralized data analysis and process on one platform like applications such as intelligent driving and video optimization. In some scenarios, such as equipment inspection and management at industrial sites, different equipment may face completely different business requirements, and the on-site data processing requirements are too fragmented. Heterogeneous data makes it difficult or impossible to provide a universal solution.

At this time, the computing storage capacity of edge computing is directly sunk to the edge of the device, that is, the device is intelligentized, and the data is processed directly at the device end, and the analysis results are directly used to guide the production of field equipment, which can solve the problem of diversified demands on the edge side, thereby reducing the production cost of the enterprise and improving the production efficiency. At present, the applications where the computing power directly sinks to the end are mainly concentrated in three areas: smart security, industrial Internet, and smart home. The common feature is that the number of terminals is huge and the data generated is many and miscellaneous. If cloud computing is adopted, either it does not meet the needs of specific business, or it will cause a lot of waste of network resources.

The core of building an enterprise industrial Internet system is platform. Industrial Internet platform is an industrial cloud platform for manufacturing industry to meet the needs of digitalization, networking, and intelligence. It builds a service system based on massive data collection, aggregation, and analysis and supports ubiquitous connection, flexible supply, and efficient distribution of manufacturing resources. The first layer of the industrial Internet platform is the edge layer. At the edge, the database of the industrial Internet platform is constructed through a wide range of deep-seated data collection.

The edge layer processes data at three levels:

1. Access various equipment, systems, and products through various communication methods to collect massive data;
2. Reliance on protocol conversion technology to achieve normalization and edge integration of heterogeneous data from multiple sources;
3. The use of edge computing devices to achieve the aggregation of the underlying data and the integration of data into the cloud platform.

Edge layer is the basis for data collection and preprocessing to be implemented in the industrial Internet. Edge intelligence is particularly important for some specific business scenarios. For example, in the textile industry, the traditional method is to inspect the defects of cloth by manual inspection. Limited by the subjective consciousness, experience, environment, cognition, and other factors of the inspectors, the inspection results are often of great difference and poor consistency. After the introduction of edge computing, the intelligent devices with simple computing ability are directly distributed on the edge of the equipment. The relevant data are collected when the machine is running, and the data are processed and analyzed directly in real time on the site, so as to minimize the product defects caused by time delay.

The application of edge computing in the industrial Internet is actually more based on the concept of intelligent transformation of edge devices. The core concepts of both are relatively similar. Both of them use the way of installing intelligent chips or external intelligent devices in the terminal to make the device have certain computing capacity, so as to realize control of the devices' own production processes and achieve the intellectualization of the factory.

For example, Jabil Circuit has been using this technology to detect errors in the early stages of circuit board manufacturing, enabling it to identify defects in the second or third step of the 35–40-step process. The accuracy of fault detection is 80%, which saves 17% of work cost and 10% of energy. When problems are previously detected, it can make the company more efficient, accelerate cycles, and improve customer satisfaction. However, this method needs 5G to access a large number of real-time and high-quality data to achieve the highest efficiency.

With modern traceability methods, such as QR code scanning and RFID, product records are usually only made on arrival, and only the location and time are recorded. If damage occurs, even though it is not impossible, determining which stage of the journey it occurs is difficult. However, next-generation wireless technology can take packaging tracking to the next level. By installing 5G sensors on the packaging, supply chain stakeholders can identify information about the packaging, such as location, temperature, humidity, gravity, and humidity. They will be able to get real-time feedback on the status and condition of the product without the need for manual checkpoints.

The production scenarios of large enterprises often involve scenarios such as cross-factory and cross-region equipment maintenance and remote problem location. The application of 5G technology in these areas can improve operation and maintenance efficiency and reduce costs. 5G brings not only the interconnection of all things but also the information interaction of all things, making the maintenance of smart factories beyond the boundaries of factories. In the future, each object in the factory will be a terminal with a unique industrial Internet logo, so that the raw materials in the production process have the "information" attribute. Raw materials are automatically produced and maintained based on "information." Raw materials, equipment, and products with a unique industrial Internet logo are recognized for information interaction. At the same time, with the help of IoTs, products and raw materials are all directly connected to all kinds of related knowledge and experience databases. When troubleshooting, we can refer to a large number of experience and expertise to improve the accuracy of problem location, so as to realize the whole life cycle management of products.

Product life cycle management refers to the management of information and processes in the entire life cycle of a product from demand, planning, design, production, distribution, operation, use, maintenance to recycling, reuse, and disposal. It is not only a technology but also a manufacturing concept. It supports concurrent design, agile manufacturing, collaborative design and manufacturing, networked manufacturing, and other advanced design and manufacturing technologies (Fig. 5.4).

The intelligence of manufacturing life cycle activities mainly includes three steps:

The first step: "feel" the industrial process and collect massive data.

The foundation of AI is based on large amounts of data, and industrial sensors can obtain multidimensional industrial data. In addition to device status information,

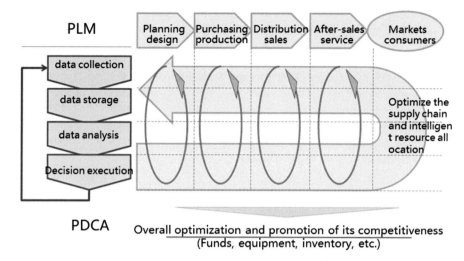

Fig. 5.4 Product life cycle management

AI platforms need to collect information about the working environment, such as temperature and humidity, to predict future trends. This requires the deployment of more categories and numbers of sensors. Nowadays, the most commonly used sensors include pressure, displacement, acceleration, angular velocity, temperature, humidity, and gas sensors. Today's industrial sensors can help to confirm the raw materials available in inventory, replace the indicator for more accurate readings, collect data in harsh environments, and monitor data transmission through the gateway.

The second step: networking – high-speed transmission, cloud computing, and interconnection.

After getting a lot of data, how to transfer it to the cloud? This requires relying on advanced industrial-grade communication technology. Different from the simple response to data directly in the workshop in the past, enterprises need to aggregate data from different workshops and factories, at different times to the same cloud data center, and perform complex data calculations to extract useful mathematical models. This puts forward new requirements for industrial communication network architecture and promotes the popularization of standardized communication protocol, such as 5G and other new technologies.

The third step: open up the data flow in each link of the supply chain, and realize the digitalization and intelligence of the entire process of the product life cycle.

The logistics between each link of the supply chain will produce a lot of data. The collection of these logistics information can help the logistics industry to improve efficiency and reduce costs. The future of intelligent logistics achieves comprehensive analysis, timely processing, and self-adjustment, through intelligent collection, integration, and processing of logistics procurement, transportation, warehousing, packaging, loading and unloading, distribution, and other information for each link. This needs to involve digitizing these data and accumulating them into a sufficient database, which requires a lot of infrastructure construction.

Smart manufacturing life cycle activities need to realize the interconnection of the entire life cycle of products from design, manufacturing to service and then to scrap, recycle, and reuse. In the future, factories will create virtual models for physical objects in a digital way to simulate their behavior in a real environment. By building and integrating the digital twin production system of the manufacturing process, it can realize the whole process intelligence from product design, production planning to manufacturing execution. It also takes product innovation, manufacturing efficiency, and effectiveness to a new level.

5.4 Enhanced Man-Machine Cooperative

The Industry 4.0 revolution is intensifying, and manufacturing elements such as machines, equipment, people, and products are no longer independent individuals. Instead, they are closely linked through the industrial IoTs to achieve a more coordinated and efficient manufacturing system.

5G has begun a new chapter in the development of mobile communications. People can watch high-definition video at anytime and anywhere; enjoy the new experience brought by VR; participate in video conferences, live sports events, and other activities; and watch movies with a VR headset, etc. 5G and VR/AR technologies can not only achieve the replication and spread of high-quality resources in medical education but also reduce the cost of medical resource flows such as telemedicine and surgery.

Increasing applications for industrial wearable devices require wide-area, large-bandwidth, and low-latency 5G network environments. In the future, people will play a more important role in the production process of smart factory. However, due to the high degree of flexibility and multifunction of future factories, this will place higher demands on the workshop staff. In order to quickly meet the needs of new tasks and production activities, AR will play a key role in the intelligent manufacturing process and can be used in the following scenarios: monitoring processes and production processes; step-by-step instructions for production tasks; and remote expert business support. In these applications, auxiliary AR facilities need maximum flexibility and portability for efficient maintenance (Fig. 5.5). Therefore, it is necessary to move the device information processing function to the cloud, as the AR device only has the function of connection and display, while it connects with cloud through a wireless network. AR device can obtain necessary information, such as production environment data, production equipment data, and troubleshooting instruction information, in real time through the network.

In this scenario, the display content of the AR glasses must be synchronized with the movement of the camera in the AR device to avoid the step out of vision range. Generally, when the response time from visual movement to AR image is less than 20 ms, it has a better synchronization. Therefore, the cloud backhaul required to

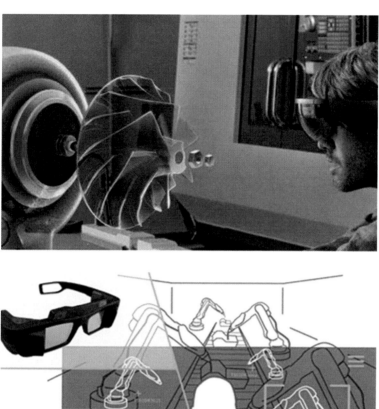

Fig. 5.5 AR human-machine collaboration implemented by eMBB technology

transfer data from the camera to cloud and to AR display content needs to be less than 20 ms. Except for the delay of screen refresh and cloud processing, the two-way transmission delay of the wireless network needs to be within 10 ms to meet the needs of real-time experience. However, the delay requirement cannot be met by the LTE network.

Flexible manufacturing is actually not a new topic. As early as the 1970s, flexible manufacturing has been proposed in Japan's FANUC machining center, where flexible unit groups for robots were implemented. But it was not until the 1980s that an automated workshop consisting of 60 units and a warehouse was built. Imagining that if you want to achieve flexible manufacturing, you need to be equipped with a large number of automated robots. The more units you have, the more machine tools you need. At the same time, you are also required to have an independent warehouse and a matching software.

This matching software has high demands. First of all, the design software must be highly adaptable. The design of drawings not only is implemented manually by CAD or NG but also requires AI for big data analysis to form an industry-level database as knowledge base, while manual work only needs a low degree of participation. But at present, the biggest problem of design is that frequent design changes lead to too many process versions. And too many process versions will lead to a higher programming repeatability.

Among these industry-level software, there is a kind of MES application. First, it can realize the process management among mold design plus process manufacturing and programming. Second, integrate machine data collection to automatically collect machine tools and equipment, improve equipment OEE (Overall Equipment Effectiveness), and present the docking ERP system, so as to realize real-time informationization of the whole processing process and be able to respond changes quickly.

Although the above methods can achieve personalized processing and realize the process of small batch products at the same time, there is a big drawback which limits the promotion of this mode. Since it is a small batch processing, it is quite likely that only small- and medium-sized processing enterprises are needed. However, to build a set of industrial-level flexible manufacturing hardware, a complete set of ERP and intelligent manufacturing, as well as software solutions such as drawing and document management, are required, and their costs may not be affordable for small- and medium-sized enterprises. To be aware of, a machine tool in the middle link of flexible manufacturing requires millions or even tens of millions.

The most important thing is that the Industry 4.0 era focuses on manufacturing itself. Because in the manufacturing process, manufacturing is the creation of value, and the production process is completed through intelligent automatic processing. What does that mean? On the industrial 4.0 platform, an intelligent design platform is built through the "industrial Internet," and the large-scale equipment owners become the processing platform, which is more centralized.

Small- and medium-sized enterprises produce value through real "manufacturing." Real production can be done by the flexible manufacturing providers of the industrial Internet platform. Of course, in this mode, the manufacturer of flexible manufacturing is not necessarily a large-scale enterprise; instead it can be a small- or medium-sized processor. As the production process is a unified standard, it eliminates half of the manual communication and can directly talk to the equipment through the product.

In the future smart factory, various production materials such as AGV and mobile robot will work in cooperation with manpower, machine equipment, and system, and positions of these machines and robots will change at any time. Therefore, providing accurate information about their locations in the factory is crucial for self-guided and efficient workflow. On the one hand, intelligent systems such as mobile robots and AGVs report their locations and time information at any time; on the other hand, related objects, such as workpieces, fixtures, AGVs, and robots, can be bound with electronic tags. The tag signal will be collected by the host system, and its position will be calculated, and then the information will be fed back to the automation system and manufacturing unit.

In the workshops of BMW and Mercedes-Benz, the car body was originally sent to different locations for processing by a single conveyor belt. Through the improvement of production technology and the application of RTLS, the fixed conveyor line of monorail transportation may be liberated. This means that the production route can be flexibly adjusted and the entire process production layout of the factory can be quickly adjusted. Therefore, flexible production will become easier and the factory space can be greatly compressed. Behind this, it needs to accurately capture the location information of each equipment and workpiece material and provide corresponding services.

The tools started to become a bit more "naughty" like a Smurf, and they can be manually operated by themselves. In the automobile processing technology, there are many tightening actions when a car is installed. It can collect the movement and its times of the tool and analyze how it can be assembled faster at a certain workstation. If some parts at this station need to be turned three times to fasten, a worker does not need to go through the operation manual or even use the memory. Instead he/she just holds the tool with positioning, and this device will automatically rotate three times. When it reaches the next process position, it may automatically rotate 15 times according to the positioning. Different tasks can be assigned at different locations. When models on the production line are changed, the installation sequence and process are changed as well. Therefore they can be easily adjusted and applied flexibly.

With the help of eMBB technology, the real-time positioning system just sends the location information to the task distribution device, which then sends the task to operating tools according to the wired or wireless communication method.

Furthermore, 5G networks enable a variety of business needs with differentiated characteristics. In large factories, different production scenarios have different requirements for network quality of service. The key to the high-precision process is time delay, and critical tasks need to ensure network reliability, real-time analysis, and high-rate processing of large flow data. With its end-to-end slicing technology, 5G networks have different quality of service in the same core network and can be flexibly adjusted as needed. For example, the reporting of device status information can be set to the highest service level.

In the next 10 years, 5G networks will cover every corner of the factory. In the process of smart factory production, it involves the judgment and decision of logistics, material loading, storage, and other schemes. 5G technology can provide an industrial Internet platform for smart factory. Precision sensor technology acts on countless sensors, reporting information status in a very short period of time, collecting industrial big data through 5G network, and forming a huge database. Industrial robots combine the supercomputing ability of cloud computing for independent learning and accurate judgment and provide the best solution. In some specific scenarios, with the help of device-to-device (D2D) technology under 5G, the direct communication between objects further reduces the end-to-end delay of the service, and the response is more agile while the network load is shunted. The time of each link of manufacturing becomes shorter, the solution is faster and more optimized, and the manufacturing efficiency is greatly improved.

Chapter 6
Postscript: Industrial 5G: Open Intelligent Manufacturing New Era

5G is ready to change the world in a variety of ways, from faster HD video to driverless cars, and telemedicine surgery through real-time response. Its unprecedented speed and coverage will bring the world closer than ever before and produces unprecedented super capabilities. It will also change many industries and including manufacturing is just an example.

5G has given common technical attributes to mobile communications and has become an indispensable and critical infrastructure alongside road networks, railway networks, and energy networks. It will comprehensively promote the development of a digital society. The typical application scenarios represented by smart cities, smart environmental protection, and smart homes are deeply integrated with mobile communications, and hundreds of millions of devices will be connected to 5G networks. 5G will also deeply affect vertical industries such as logistics and construction with its ultra-high reliability and ultra-low latency performance. For example, 5G technology in Industry 4.0 can help smart factories achieve autonomous decision to improve operation and maintenance efficiency, reduce costs, flexibly produce diversified products, and quickly respond to market changes.

The manufacturing industry is a highly complex industry. A product has as few as tens of raw material inputs and is made up of as many as millions of parts. To produce the same product, different companies have different production processes, production equipment, and parts inputs. Due to different production processes, equipment interfaces, and data formats, not only will the digital connection of the upstream and downstream of the supply chain be difficult, but also the digital transformation of each enterprise will have to be a new start, which is time-consuming and laborious. Based on the 5G communication infrastructure, building a more versatile and plug-and-play industrial Internet platform that follows common standards can solve the abovementioned problems in an intelligent manufacturing process.

It can be said that 5G technology is bringing Industry 4.0 into another stage of development, which will affect the development of applications from aspects

X. Wang, L. Gao, *When 5G Meets Industry 4.0*,
https://doi.org/10.1007/978-981-15-6732-2_6

such as connectivity, security, and computing. The biggest change in 5G is to enrich the scope of network connection through the definition of three major application scenarios, thereby meeting the new connection needs and expanding the network connection from "people" to "things." If 4G changes the way people connect to each other, then 5G has changed the multiple connections methods of "people and things" and "things and things," and realize the digital connection and high degree of collaboration of production equipment, value chain, and supply chain. It enables production systems to have the capabilities of agile perception, real-time analysis, autonomous decision-making, accurate execution, and learning enhancement. It comprehensively improves production efficiency, enables rapid upgrading of engineering operations and manufacturing automation technology, and enables large-scale manufacturing enterprises to achieve more intelligent manufacturing processes. It reduces the demand for production line workers, accelerates the development of industrial intelligence and automation, and reduces the demand for traditional labor-intensive work. At the same time, through the 5G network, it enables real-time monitoring of product status and responding to user needs; provides value-added services such as rent-on-sale, on-time billing, remote diagnosis, failure prediction, and remote maintenance; and achieves the high additional transformation of manufacturing enterprises from providing products to providing "products + services". In addition, 5G will also be used more to improve quality control, collect various data in real time with low latency through uRLLC technology, and train data sets according to AI-based visual inspection systems to ensure that they can identify all potential defects, so that workers can quickly identify errors and mistakes that weaken the production chain and product quality.

When 5G meets Industry 4.0, CPS will be closely linked. That is, IoTs will connect processors and sensors at the production site and allows robots to communicate and talk with each other. Therefore, works of machines and humans will no longer be strictly divided. In the future, manufacturing systems will integrate people and machines, by using the algorithm models such as machine learning, pattern recognition and cognitive analysis, to not only enhance the capabilities of factory control and management systems but also achieve the so-called smart manufacturing. So that enterprises can obtain better advantages in today's competitive environment.

We are convinced that the digital transformation in 5G's fast-moving vertical industry and the huge user scale will promote the rapid and healthy development of Industry 4.0. Industry 5G will be a super key to open the door of smart factories. In the 5G era, people have more imagination for smart factories, and those invisible data have freed themselves from the shackles of equipment and moved from the dark of the machine to the bright screen. And industry 5G, with this traditional role, is commanding the free flow of all data in the background with unprecedented inclusiveness.

Printed in the United States
by Baker & Taylor Publisher Services